# INTUITOUCH

# INTUITOUCH

*Healing through the Gift of Intuition and the Art of Touch*

Dr. Jim Bourque Starr

With a Foreword by

Dr. John F. Demartini

authorHOUSE®

AuthorHouse™
1663 Liberty Drive
Bloomington, IN 47403
www.authorhouse.com
Phone: 1-800-839-8640

First Edition 05/20/2005

First published by AuthorHouse    09/08/2011

ISBN: 978-1-4634-2044-4 (sc)
ISBN: 978-1-4634-2043-7 (ebk)

Library of Congress Control Number: 2011909599

Printed in the United States of America

Any people depicted in stock imagery provided by Thinkstock are models, and such images are being used for illustrative purposes only.
Certain stock imagery © Thinkstock.

This book is printed on acid-free paper.

# TABLE OF CONTENT

# FOREWARD

What if you could transform your life and manifest the life you would truly love? And what if there were a new way of unveiling your true magnificence? Well, now there is. Though most people have relied heavily on their five primary senses for their survival and existence, some individuals, who have been blessed by becoming more awakened, have learned to listen to their sixth sense. Dr. Jim Bourque Starr is one such gifted individual. It is this enlightening inner sense, along with its corresponding sixth expression—the "*InTuiTouch*"—that opens the doorway to a new and more empowered state of well-being and heartfelt living. The special book that follows, purposefully entitled *InTuiTouch*, blends the inner wisdom of this extra-special sense with the magical powers of hands-on healing. Ever greater inner healing methods for the future are now emerging. The frontier of the modern wellness methods is concentrating on the subtle energy vibrations of the mind and heart.

Realizations about the influence of our minds and emotions on our bodies are rising in the media, and awareness of Eastern and Western healing approaches is certainly becoming integrated. The "*InTuiTouch*" concept blends the essences of these emerging wellness approaches and distills them into a higher-ordered method that can—and has—transform lives and heal bodies, minds, and souls. It represents a gift for humanity that will spread and touch millions of individuals over the years to come.

As you read this magical book, contemplate the very message that each chapter brings; sip from the chalice of wisdom that they hold and reveal. May your journey now be enhanced, uplifted, and enlightened, and may your body, mind, and soul be raised to new levels of vibration so as to express your divinely designed magnificence.

*Dr. John F. Demartini*

Best-selling author of *Count Your Blessings—the Healing Power of Gratitude and Love* and *The Breakthrough Experience—A Revolutionary New Approach to Personal Transformation.*

# ACKNOWLEDGMENTS

Like many beginning authors, I am here to tell you that you can't even imagine the number of times that I have begun to write this book. How often can one make a personal commitment to oneself—only to break that commitment? What you have in your hands is the culmination of all of those broken commitments. It took a reason, a passion, a purpose, and, more than anything else, it took the love in my life to help push me over the top and finally finish what I had set out to do.

I dedicate this book first and foremost to Araceli. Just as her name implies (guardian of the heavens), Araceli has been the guardian of my mission. The dominant color in her aura photo is violet, which indicates that her essence is her spirituality and that she is both a wizard and one who manifests her dreams. The truth is, if it were not for her, this project would probably still be in the planning stages. She pushed me to complete my work when my distractions tried to pull me in the opposite direction. She is my best friend. Thank you, Araceli I love you with all my heart.

My second round of thanks, but certainly no less significant, is to my children: Jimmy, Amanda, Nicole, and Kelsey. They have received enough challenges and support during all of this to last their lifetime. It has been said that "greatness is measured by the trials and tribulations as well as the responsibilities placed upon one's shoulders." I am so proud of the way my children have responded to the life that has been presented to them.

Within each page of this book lies a part of them that has contributed to my life, contributed to making this book possible. They are of the highest value in my life, and I have paid an immeasurable price to be without them during the countless hours of preparation of this book.

My third round of gratitude must not be overlooked, and it goes to my beautiful, loving sister Valerie. I don't know why she hasn't written her own book yet. Her wisdom and light need to be shared with the world. Know for a fact that Val's influence is here, on every page. She believed in me when even I didn't. She was there for me during the most difficult times of my life.

The trials and tribulations I experienced as a child of alcoholic, angry, confused, unhealthy parents were the greatest lessons; they drove me into preventative health care and ultimately into my metaphysical studies. No matter what, I have always known that you, Valerie, will always be there for me.

What more can I say to Araceli, my four beautiful children, and Val except thank you, and I love you.

# INTRODUCTION

# INTRODUCTION

W hen you talk to God, you call that "prayer." When God talks to you, it is called "intuition."

The art of listening and paying attention to my "gut instincts" has led me to the development of a unique approach to healing called "*In Tui Touch*." Please open your heart and listen by feeling. It will greatly impact your health, and it may even save your life.

Throughout this book, you will find short anecdotes. I have called them "*Real-life Stories.*" Although the names have been changed to "protect the innocent," I can assure you the stories are real and have made such an impact in both my professional and personal life that I felt compelled to share them with you. I hope that you enjoy the significance of the messages that they bring so that you can apply them to your own life situations. There are no coincidences in life, and when one or more of these stories hits you like a ton of bricks, know that God is at work.

## REAL-LIFE STORY

*Linda*

She was thirteen and only wanted what all thirteen-year-olds want: peer acceptance (fitting in) and a clearer sense of who they really are. It appeared, however, that God had other plans. Her rapidly failing eyesight forced her to wear ever-thicker lenses. With each ensuing change to a stronger prescription, Linda's self-esteem sank deeper and deeper. It was obvious in her slumping posture and her introverted personality that she was not a "happy camper." The more she withdrew from the world, the less peer acceptance she had.

The best eye specialists in the area had no explanation for her failing vision although they concluded that within the next two years, Linda would most likely be considered legally blind. Her father, Roberto, scheduled an appointment with me as a last-ditch effort, hoping that chiropractic adjustments could somehow make a difference. It was only my rookie year in practice, and I was quite reluctant to see Linda. My only justification for attempting chiropractic (other than a need to see new patients) was the fact that Linda had been in a car accident precisely at the time that her vision began to fail. The possibility existed that there could be a connection between her failing vision and nerve pressure caused from misalignment of the spine that had resulted from the accident. The "possibility" is all that Roberto and Linda needed to begin a course of treatments. After all, what did they have to lose?

What transpired over the course of the next three months not only dramatically changed Linda's life with the restoration of her vision, but it completely changed my life as well. Just as the medical specialists said there was no explanation for her *failing* vision, now they had no medical explanation for her *recovery*.

Spontaneous? Miraculous? It was outright unexplainable. All that I know is that there was something bigger than just my chiropractic treatment going on here.

From the onset of treatment, I established a great doctor-patient relationship with this thirteen-year-old girl. I had a gut feeling and just "knew" that there was more involved in this scenario than a pinched nerve. Something as large as Linda's life was involved in this imbalance in her vision. Naturally, she received the chiropractic adjustments to her neck because, on a physical level, there was much work to be done to realign her spine. But something inside me kept drawing me closer to look for a deeper cause than simply the obvious misaligned vertebra. I knew that if I could help her "see" the hidden cause, she would begin a real process of healing. Because of our deep trust and faith in what we were doing (the doctor-patient relationship), Linda also had a deep sense of certainty that she would heal. Our unified certainty helped us discover the "what" and the "why" that she didn't want to "see."

Linda's confidence in me allowed her to reveal that her mother had broken the news of a pending separation and divorce from Linda's father. The news coincidentally had occurred the same day as, just prior to, the car accident. When Linda had heard the news, fear, guilt, and anxiety ran through her blood like fire. She couldn't and wouldn't accept it. She put this nightmare into the closet of her mind and didn't want to even look at it. With this newly shared information, our awareness of the mind-body connection was very evident. Her trust in my approach allowed our energetic connection to flow in a safe, harmonious, and powerful way.

With her adjustments to her neck (the physical imbalance) and the acknowledgement of what she didn't want to see, miraculously, Linda began to heal. Her eyesight changed. The week-to-week improvement further solidified our certainty of the outcome. The steady improvement not only changed her posture physically, but it also changed her posture in her life. Her self-esteem began to soar. Her friends noticed the difference. Her medical doctor was scratching his head in disbelief by this time.

Three months later, Linda's eyesight was 20/20. The experience that I had with Linda opened my heart as well as my mind to a world that went far beyond my medical training. I realized after that experience that science and logic serve in healing—but in a limited capacity. There was much more to healing than a pill, potion, surgery, or a chiropractic adjustment to align the spine. There was divine intervention.

It was through that experience that my intuition (God talking to me) led me to a feeling, which I listened to and didn't ignore. This led me on a quest for answers for the next twenty-five years that changed the course of my medical career. Since that experience, I have been on a quest to learn and develop all of my intuitive abilities as a healer, never to deny them again.

How many times have you received signs and simply ignored them? If you are like I was, you can't even remember how many times. Those gut instincts can and should be trusted. Only through faith, practice, and experience will your awareness of signs serve you. We can pray for signs (us talking to God), but the most important part is the ability to *listen* to when God is talking to us.

This is the gift of intuition. I call it a gift, even though we all have it, but, like any gift, what you don't use, you lose. What happens if you are given an apple as a gift and you don't eat it? It will spoil and rot. This book will help you get in touch with your intuitive abilities so they don't rot and lie dormant. It will help guide you not only in your health, but also in your whole life. Intuition has even served as a catalyst for some people to discover their hidden talents as a healer—enabling them to help others as well.

We are in an age of conscious evolution far faster than any period in the history of civilization. We are quite aware of the energies that exist beyond the five senses. Listen to your heart. *InTuiTouch* is part of this evolution, and I am proud to bring it to you.

# CHAPTER ONE
## WHY INTUITOUCH?

# CHAPTER ONE

# WHY INTUITOUCH?

A famous author and philosopher once said, "If the why is big enough, the what and the how will take care of themselves." The why in my life was birthed out of my frustration as a healer. It seemed that, whatever treatment a patient received, regardless of the discipline, the results were only temporary. The healing arts abound with treatment options: Reiki, polarity, reflexology, homeopathy, acupuncture, chiropractic, naturopathy, and many, many others. Each of these techniques approaches the same health challenge in a different manner; all serve people well. Some balance energy, some balance emotions, and some balance posture. Some of these approaches address the body, some the emotions, some the mind, while others address the spirit. There are even some techniques that address a combination, such as mind and body. I recognized that something in the healing formula was missing—and that something was a reproducible technique that actually addressed the mind, body, emotions, and spirit simultaneously.

*This is what my intuition told me,* and it led me to two conclusions regarding health and healing.

**Conclusion #1:** Maximum healing occurs when all four dimensions of our physical, mental, emotional, and spiritual being are aligned.

**Conclusion #2:** Every health challenge, regardless of its name or its severity, involves all four dimensions of our being simultaneously.

Ever since my experience with Linda, I have been searching for answers. I knew that there was more to "that shoulder problem" than simply stress or tightness; I knew that the perception of "carrying the weight of the world on one's shoulders" (responsibility or guilt) was a deeper, core issue. With many of my earliest patients, I felt as if I couldn't talk about these core issues because many people are too uncomfortable with the truth. It seemed that many of them, who were used to traditional chiropractic, simply wanted the adjustment so they could go home and feel better for a while. Time and time again I would find myself rationalizing, "at least they're not taking pain medications for the relief of their pain." For me, however, it really wasn't enough. I knew *intuitively* that I wasn't doing all that I could do for the patients, and that, soon enough, the patient would return with similar pains. That's how Band-Aid approaches to treating symptoms works. It stops the bleeding, but it doesn't address the cause of the cut.

When I began to treat the patients in four dimensions *who were ready* to hear the truth, I found not only that improvement was rapid, but also that change seemed to occur on a larger scale. Not only did pain levels diminish; lives improved as well. Careers boomed, relationships deepened, gratitude returned, family connections improved. Patients improved because the "why" of the health challenge was being addressed.

Soon I faced a life-changing question regarding my future as a healer: Do I dare use only the standard treatment for patients based on their diagnosis and the conventional acceptable wisdom of care, knowing full well that they need and deserve more? Or could I stand firm in what I knew was true and trust that I would attract those patients who were ready for this? I decided to trust my intuition again.

At some point in my early career, I decided that the truth would set me free, and that is when I began a new chapter in my life as a healer. I needed to utilize a *complete* technique, one that addressed the physical, mental, emotional, and yes, the spiritual part of who we are. Whether I encountered resistance didn't matter to me anymore because I was practicing the Truth . . . *my* Truth.

The development of the **InTuiTouch** Method came about as a result of my need to have a simple, systematic, and yet complete way to address any and all health challenges. No more piecemeal work; I would address four dimensions simultaneously. What is so beautiful about the **InTuiTouch** Method is that it integrates so well with all techniques and all disciplines of healing. This is so because the fundamental key for patients is acknowledging the four dimensions and then utilizing their own intuition. This will to lead them to the answers that will help them on the path to healing.

Intuition is a talent that we all possess. It is as natural a gift as anything about us. Learning to use intuition in healing either as a healer or as a "healee" simply takes practice—like anything else in life that has any real value. Think about the things that mean most in your life and that you are the most proud of. Are they the things that you sacrificed most for and that were the hardest to achieve? Intuition is not hard to achieve, but it does take time to refine the gift that we already possess.

Welcome to the place where the gift of intuition plus the healing art of touch meet: ***InTuiTouch***. This is where the process of healing is *complete*. It is a gift for the world.

# CHAPTER TWO

## HEALING IS AN "INSIDE JOB"

# CHAPTER TWO

## HEALING IS AN "INSIDE JOB"

E very time people stop to think about their health, they stop and analyze whether or not they have any symptoms at that time. This is rather interesting, Dorland's Medical Dictionary, a widely accepted "bible" of medical dictionaries, states, "Health is that state of optimum physical, mental, emotional, and social well-being, not merely the absence of symptoms in the body." In other words, pain or symptoms do not necessarily determine whether one is sick, and absence of pain or symptoms does not necessarily determine whether one is healthy.

## REAL-LIFE STORY

Even though the lobster tail had a "curious" smell to it, Ron didn't want to complain on his first date with his new acquaintance, so he ate it anyway. Besides the macho pride factor, the onions, garlic, butter, and spices seemed to mask any foul taste. Ron overlooked some obvious signs of danger, dismissing it as his imagination. But his romantic evening was cut short by an intense and disappointing stomachache. One hour later, his nausea was followed by fever and violent vomiting. Ultimately, diarrhea set in.

Reviewing Ron's symptoms, we note: stomachache, nausea, fever, vomiting, and diarrhea. Obviously, Ron was very sick, right? Wrong. Ron is healthy. Thank God, Ron had these *symptoms*. Ron had ingested a poisonous, spoiled lobster tail that was full of harmful bacteria that, if left inside of his digestive tract, could have killed him. The bacteria could have perforated his bowel and caused septicemia or any of a dozen other complications. Consciously, Ron was not aware of what danger had entered his body. Unconsciously, however, his inner diagnostician was busy detecting and protecting the body from harmful invaders, and it immediately began working to get rid of them.

We call these natural reactions of the body . . . symptoms. In our medically dominated society, stomachaches, nausea, fever, and diarrhea are looked upon as "bad" and "sickly" but, in truth, they are the healthy reactions of our bodily functions and were designed to help preserve life.

Now, let's examine how these healthy reactions are looked upon from an allopathic standpoint, that is, the viewpoint of "conventional" Western medicine. We have at our disposal manmade chemicals designed to eliminate each of these symptoms from our body—symptoms that the body is creating in an effort to eliminate the toxins. For stomachache, we have "plop, plop, fizz, fizz." For the fever, we have of course, our aspirin (80 million taken per day in the United States alone). You know the names of the pink anti-diarrhea and anti-vomiting, drugs that we are urged by TV ads to take every time we have these symptoms.

If Ron were to take these chemicals to mask the symptoms, as advised by the TV ads, he would actually be preventing his body from expelling the harmful bacterial agents. He could, in fact, induce a much more serious problem by keeping these harmful agents inside to do their damage. Many times, "modern medicine" is more dangerous than the condition at hand. Statistics show that, last year alone, more than 100,000 deaths were caused by adverse drug reactions (ADRs) that occurred in our hospitals, with patients who took the *correct* dosage of the *correct* medicine for the *correct* diagnosis! That fatality number is equivalent to having three jumbo jet crashes every two days for one year. Yet, we rarely—if ever—hear about the number of deaths caused by this medical tragedy in the newspapers or TV. If we did, people would be running out of the hospitals like herds of buffalo fleeing hunters. Panic would set in if all of the statistics were revealed about our so-called scientific medical approach to health. We certainly hear about the carelessness of our airline industry if even one jumbo jet crashes. Some call this the "protection of big business." You can draw your own conclusions.

The natural approach to Ron's health challenge would be to acknowledge that his symptoms were signals from his body that something was wrong and then allow the body to continue to do what it was designed to do—eliminate toxins that are harmful to the body.

*The more we are inherently aware of our body's signals, the healthier we are.*

# THE SILVER LINING IN THE DARK MEDICAL CLOUD

Just before the turn of this century, a statistical medical phenomenon occurred that revealed a change in the consciousness of our society.

According to the National Institute of Health (NIH), there were *more* office visits by US citizens to non-medical alternative health care practitioners than visits to allopathic medical doctors for the *same* health conditions. The magnitude of this change in patient preference clearly shows that modern medicine is losing its stronghold on the "business" of medicine. It is no wonder that the AMA had attempted to advocate for legislation that would that would allow only an MD to dispense vitamins and herbs. Think about *that* one for a while!

The shift in consumer consciousness is radically changing our health care delivery system. We now have hospitals competing for business by touting their "natural birthing centers."

My question to you is, since when is giving birth *not* natural? As a member of the health care industry, I cannot overlook the value of modern medicine in an emergency, in life-threatening situations that many people in our society face. I have the deepest respect for the emergency teams, surgical wards, and lifesaving drugs developed by my fellow doctors. By comparison though, the number of drugs used to calm the patients' everyday symptoms is astronomical and out of control. These synthetic chemicals that are used to reduce or eliminate our natural symptoms and signs carry with them an average of ten or more side effects that are potentially more harmful than the condition they are designed to treat. It is my opinion that a lot more thought and consideration should be given to the decision to take these drugs.

There is no doubt that more and more people are seeking conservative, natural methods of treatment before resorting to drugs. That is a silver lining beneath the cloud. We are getting the message. Our influence is sending a strong message to the drug industry to acknowledge the side effects of drugs and drug interactions. Our society is returning to Mother Nature every chance we can.

A famous man was once quoted as saying, "The doctor of the future will give no medicine but will interest his patients in care of the human frame, in diet, and the prevention of disease." Sounds pretty logical, doesn't it? It even sounds like one of those TV nutritionists that come on every Sunday. This quote, however, is by a man who was way ahead of his time. This was said by Thomas Alva Edison, just before he died.

Think about where health care is today and where it is finally going. Every drug has a list of side effects. The labels of our foods have nutritional information. Thirty years ago, to ask for a second opinion after visiting your medical doctor was considered shameful and disrespectful; now it is the norm and not the exception.

Alternative health care providers like chiropractors, acupuncturists, naturopaths, and even energy healers are commonplace, and these treatments are even covered by some insurance companies. This was unheard-of until recently. Why the change? Certainly it is not because of politics. The AMA remains one of the strongest political forces in Washington. So are the drug companies and the hospitals. The change is coming from *societal pressure* and a shift in health consciousness.

We are recognizing that health is an inside job! It does not come from a knife, potion, pill, or needle. *While medicine helps our bodies in a crisis situation, it is only Mother Nature that heals—and she heals from within.* Even the surgeons recognize this fact when they say, "The surgery was successful; now let's see how the body responds," or "take these pills and let's see how your body reacts."

# CHAPTER THREE

## GOING WITH THE FLOW

# CHAPTER THREE

# GOING WITH THE FLOW

The more the doctors and the healers work with Mother Nature, respecting the healing process, the more likely it is that a favorable, long-lasting healing will occur. We all have been blessed with the capacity to create healing within ourselves and others through Mother Nature because we are part of Mother Nature. The more in tune we are with the flow of our body's signals, the healthier we remain. Go with the flow. We cannot avoid sickness; it is part of Mother Nature's homeostatic response to the environment and our own state of mind, body, emotion, and spirit. We are dynamic beings, and, therefore, we are ever-changing. We will continue to have challenges and support from our physical environment as well as our emotional environment. To resist this, to block it out or ignore it, will ultimately lead imbalances in our system and other consequences. Imbalance is equal to disease or illness. As humans, we tend to run away from pain or challenges, suffering and illness. The natural laws of balance, cause and effect, equilibration, yin and yang, and homeostasis mean that no one can avoid or escape these events. We can, however, maintain optimum states of health by maximizing our connection with nature via the foods we eat, the air that we breathe, the exercise that we do, and the actions we take to minimize stressors. In the same light, we can jeopardize our health status by doing the opposite: eating the "wrong" foods in excess, avoiding exercise like it was the "plague," and allowing ourselves to accumulate more stress than necessary via worrying. Going with Mother Nature and the flow serves us in so many ways.

Take this information to heart. Understand that symptoms are the signals that help us understand the body and to help our bodies maintain balance or homeostasis. Nature does this on her own, as long as we do not interfere. When we alter nature, we alter the natural healing process; this puts our bodies at risk. The body is unable to perform the way it was "divinely designed" to perform. Cancer, heart attacks, diabetes, and arthritis are all indicators that something very unnatural is going on. Where did it begin?

A more important question is: Where will it end up? We have no control or guarantee of the outcome. We only have the knowledge that Nature will do its best to return to a homeostatic state of balance whenever possible. If our body cannot return due to that state due to manmade forces, it will self-destruct at a *faster* rate than it was naturally designed to do. This we do know to be true. Nonetheless, *every human body will naturally self-destruct by Divine Design anyway.*

That is the natural process of every living organism, not just the human organism. Every living organism will go through the four stages of life: birth, growth, decay, and death. From the moment we are born, we are actually going through the stages that lead to eventual death. Although this process is totally natural, humans have learned to fear the ultimate outcome, and we constantly try to prolong life unnaturally. In doing so, we promote fear of the dying process rather than welcoming it and honoring it as many of the Indian and earlier civilizations did. Ultimately, we don't escape it anyway. While prolonging life is also honored and desired, eliminating the fear of death would go a long way toward learning to value the quality of the entire process, thus reducing our stress factors and contributing to the natural homeostatic process. A wiser approach do our wellness is to live each day in accordance with nature.

Respecting life and death simultaneously is the new mantra of the modern world. This is not a new idea. Some of humanity's oldest civilizations, such as the American Indians, the Egyptians, the Japanese, and many others, since their beginnings, have honored the passing of a soul to the next world as much as they honor all living beings in this world.

Learn to maintain the balance in your life in accordance to nature. Listen to and watch the animal kingdom. Follow the philosophy of using only what you need and sharing the rest. This is contradictory to the popular notion that "more is better, especially if you have it and they don't." It is wise to eat what Mother Nature has prepared, and not what is synthetically prepared by our own mothers.

# CHAPTER FOUR

## THREE STEPS TO LONGEVITY

# CHAPTER FOUR

# THREE STEPS TO LONGEVITY

The Centenarians of the world, those who have the honor of living more than a hundred years, in following their path to health and longevity, have discovered a secret containing three common ingredients:

## *Rhythm, Simplicity, and Moderation*

**Rhythm:** When we follow a routine with our habits, such as sleeping, eating, and exercising at the same time, our bodies work harmoniously. These habits create patterns that our physical machine, the body, enjoys and benefits from. Pay attention to what times of the day or night that you have the most energy and what times of the day that you have the least amount of energy. Like all living organisms, our bodies will tell us which times work best for which activities, through what is called our *bio-energetic rhythm or Bio-Rhythm.* Even at the cellular level, our nutritional intake and waste output works best when the digestive tract dissolves the nutrients and passes them along to the circulatory system where the body absorbs them if this occurs in a *routine fashion.*

Have you ever eaten a steak dinner at 2 a.m.? Unless that is your normal routine, you will notice that you will probably incur insomnia, belching, gases, and maybe a stomachache throughout the night. The entire next day, you might feel "off," simply because of the interruption of your body's normal rhythm. The same experience holds true when you stay up too late at night or get up too earlier than normal in the morning.

**Simplicity:** The deep, rich, high-calorie French sauces that taste so incredible also put an incredible amount of stress on the digestive tract. While everyone is entitled to these fine delicacies, they should be kept to a minimum. Keep your regular diet as simple as possible, just as the centenarians have done. Choose fresh, raw foods; eat less of foods that need to be cooked; and always eat foods free from additives, preservatives; and avoid the use of synthetic cooking tools, such as aluminum pots,

Teflon coatings, etc. Simplicity should be maintained not only in the foods that we eat but in our social activities as well, by not over-scheduling or creating complications with commitments. Even the avoidance of gossip assists in the simplicity needed to enjoy longevity. Practice the law of "detachment." This law refers to the ability to be happy with what you have without feeling that you continuously have to "keep up with the Joneses." The more we give away, the lighter we feel. This feeling is very liberating, especially for those who have always been "savers."

Try, for starters, to give away a large portion of your wardrobe. You know—the clothes that you never wear but always save for "someday." In all that we do, remember the KISS principle: Keep it simple, silly.

**Moderation:** "Too much of" or "not enough of" are both signals of imbalances in life. Remember, the healthiest approach to life and living is moderation. Excess in coffee, meat, bread, or even water can cause bodily harm. Excess work or exercise will do damage, as will not enough. Too much talking, too many friends, too much praying—or not enough of any of these—is unbalanced and will cause your body to be physically stressed. Life should be balanced in the seven areas: physical, mental, spiritual, familial, social, vocational, and financial.
Trying to maintain this delicate balance will add years to your life and life to your years.

So remember the Big Three: rhythm, simplicity, and moderation. These are your major keys to health and longevity. Just ask those who have lived over a hundred years. After all, who can offer better advice on living than the ones who are "living proof"?
Only mankind strays from the natural balance. Only man loses touch with nature. The healer and doctor within will provide you with exactly what you need when you go with the flow, not against it. One of the simplest songs in the English language can help direct your life along the natural course. It goes like this:

"Row, row, row *your* boat (not what others tell you to do) gently *down* the stream (not against the current) merrily, merrily, merrily, merrily(love, happiness, humor, light)
Life is but a dream (you can create your own happiness and reality)."

# CHAPTER FIVE

## MISSION IMPOSSIBLE OR . . . MISSION I.M. POSSIBLE?

# CHAPTER FIVE

## MISSION IMPOSSIBLE?
## OR . . . MISSION I.M. POSSIBLE?

Webster's dictionary defines the word "impossible" as something that cannot be or cannot be done. Naturally, the word "possible" means something that "can be" or "can be done." Quite interestingly, between the two words, impossible and possible, lies a difference of only two letters: I and M. These letters can also represent two words "I" (me), and M (am). I am comes from the verb "to be." Quite literally, what stands between impossible and possible is the ability to be. Freedom from judgment of self and others, no worrying about what may occur in the future (fears) or concerns about what you may have done or not have done in the past (guilt). Simply *Be* . . . be right here and right now.

Neale Donald Walsch, the author of the best-selling trilogy, *Conversations with God*, states that we are not "human *doings*, . . . we are human *beings*." Stop trying to change the world, self, or others (the act of doing)—and instead, simply *love* the world, others, and yourself (the act of being).

God, who may be expressed in many forms, could be affectionately known as the "*Grand Organizing Designer*." This designer of the universe does not make mistakes. There is a hidden order of things in this universe that you may perceive as chaos, disorder, pain, and harm. From a larger perspective, however, you see (consciously or unconsciously) that everything has a reason and that there are no coincidences in life—only lessons that teach you about loving unconditionally. Your emotions might cause you to keep misperceiving that there is a disorder in your life; this might cause you to react to the world through the emotions of fear, anger, desire, and love. When someone expresses anger, there is an underlying misperception called fear. Behind this mask of fear, there is an underlying emotion of desire, which, in turn, is only masking your innate reason for being on this planet in the first place—and that is to *love*. All of the masters before us, all of the sages have been trying to teach us this principle. Jesus Christ is quoted as saying, "I am Love, I am the Light, I am the Way." Love, light, and way are all expressing the same thing: "Love can move mountains;" "Love can conquer all." By being in a state of "love," the "impossible" becomes the "possible." The miracles performed by Jesus Christ were the result of his state of "being" in the *present* (not in the past or the future); that state of "simply being" is being in the state of love. Christ is quoted as saying, "All that I have done, so can you do—and even greater things."

Is that quote clear enough? As a healer, doctor, or patient, healing begins by recognizing the loving role that God plays in your life, teaching you how to love and be in the present, to simply *be*. If it is true that God is in all things, then God is in illness and sickness as much as God is in health. If God is in everything and *is* everything, then God has created the answers or cures to these illnesses and diseases as well, even when your limited mind perceives some diseases as "incurable" (which means literally that "the cure is within"). Whatever illness or challenge you are now faced with, even if you perceive it as impossible, remember that you are in a state of fear, and behind it ultimately, is love—which is God, which is in *you*.

All is possible. The healing of cancer and other life-threatening diseases depends on changing your *state of perception* or your *state of consciousness* to a state of *certainty*.

You must go from a state of *doing*—searching for a cure, taking medications, worrying, crying, panicking, playing the role of "victim," and expecting someone or something from the *outside* (surgery or drugs) to cure you, to a state of *being*—of love and gratitude looking *within* (the kingdom of God is within) and acknowledging this power, this perfection just as it *is* (the state of "I am").

What is the "Mission" in the "Mission Impossible"?

It is about understanding that *all is possible through you*. The next question is: What mission are we talking about?

The four cardinal questions of life are: Who am I? Where did I come from? Where am I going? and, ultimately, Why am I here? Since the dawn of time, man has asked these questions.

***What I am is God's gift to me—but what I do with my life is my gift to Him.***

Everyone has a mission, a purpose, or a chief aim in life. When you are *present* (being), you are serving your mission or calling, and you get a feeling of fulfillment, energy, and satisfaction from your giving (service) and contributing. We have all experienced that feeling at one time or another. Being very present allows time to stand still. We are "in the moment," or "in the Light."

The Light is an emission from God that allows us to fulfill our mission. Plato said, "What comes to us from God looks back to God from us." The more that we adhere to being present and answering our calling in life, the more Light we receive, the more God reflects in us.

Health, balance, moderation, and love are all reflections or emissions of the Light that confirm that you are doing your mission, through submission to your Higher Power and the omission of your ego. At the very moment that you are in tune with your mission, you are healing—and everything is possible because with God, all things are possible.

When was the last time that you felt like you said the right thing to the right person at the right time changed their mood or their life? Was it at work? At home? With friends? Think for a moment; how did that feel? Did it feel overwhelmingly peaceful and satisfying? What you have just recalled is a glimpse of the gifts and talents that lead you toward your mission. The more you focus on your gift (which is called a present), the more in tune you are, and, clearly the more you are on your mission. The more that you are on your mission, the more things that you want to be, do, and have are possible.

In the emotional phase of the ***InTuiTouch*** Method, I integrate ***The Consciousness of the Heart*** transformational course. The end of the two-day course focuses on getting you to be clear about your mission or calling in life, which is an essential cornerstone to understanding the "why am I here" question. By clarifying your mission, the parts of the puzzle that seem so challenging and frightening all begin to make sense. The logic that surfaces allows you to be grateful for the Divine Design. This gratitude will open your heart to loving your state of being or condition exactly as it is. This state of love and gratitude opens your heart to more of God's love and light—and ultimately to healing. So the mission is to *be*—to simply *be*, all that God has planned for you.

Be in the present, the "*I am*" state, and all things are possible. Any financial challenge, health challenge, or relationship challenge will dissolve into an appreciation and love state that will by Divine Design, play out its verse in your song of life. Welcome it. Love it. Remain detached to the outcome.

Just simply *be*. At that moment you will be serving yourself, God, and all of humanity. Remember, with God, all is possible. The Mission Impossible becomes the Mission I.M. Possible.

# CHAPTER SIX

## INTUITION,
## THE NEW ERA IN HEALING

# CHAPTER SIX

## INTUITION, THE NEW ERA IN HEALING

Pay attention to your "gut feelings"—those sensations that you get seemingly out of nowhere. At one time or another, everyone feels them. These feelings or signs are exactly what intuition is all about.

I began using my intuitive abilities in my early childhood. I remember avoiding going with my bicycle down Crestview Drive one day, even when all of my friends decided to go. My hunch was right; I was the only kid on the block who didn't get beaten up that day by the neighborhood bullies—who were waiting beyond the trees to ambush them when they least expected it. It seemed that I was the only kid who knew where to place the crab traps when my friends and I would go out to the backwoods of Patcong Creek in South Jersey. They always said it was "luck," but I knew better. I would always go with my "feeling." I didn't realize until I was in my second year at Chiropractic College that what my mother had told me was true—when I was eight years old, she had boldly stated, "Jimmy, you have hands that are blessed."

It seemed that all I had to do was place my hands on her neck or shoulders, and her headaches and tension would disappear. I was intuitively drawn to healing, even at that young age—although at that time, most of it was unconscious. Utilizing my sixth sense came easily. My sister Val and I talked about her ability to "see" things long before they happened. She was especially attracted to the Ouija board and played the clairvoyant with the crystal ball. My mother needed nothing but dreams; her dreams gave her insights into the not-too-distant future, which she used to help guide and protect us. This blessing was also a curse; sometimes she was anxious due to "knowing" what was about to happen to family members long before it happened. Her visions proved accurate too many times to call it coincidence. The worst visions foretold illnesses and tragedies. She only told us about some of her visions. In the fifties and sixties, any talk of other worlds landed you on the couch of the neighborhood psychologist or psychiatrist—which is exactly what happened to Mom. She was prescribed

Valium and Librium just to "ground" her. The "spirits" of alcohol that she ingested between pills kept her in quite a zombie-like state until she finally decided she'd had enough.

Mom then went on to dominate her newfound refuge—Alcoholics Anonymous—for the next twenty-five years, helping hundreds of people through recovery, until her death. Once she was clear of the drugs and alcohol, Mom's visions were stronger than ever. She had learned the difference between a misdiagnosis and a misunderstanding. She was very careful in choosing whom to share her intuitive gifts with from then on.

The one person who always understood and supported the validity of Mom's talents was "grand-mom." Up to her death at age a hundred and one, people from miles around would visit with grand-mom to ask her what their future held. She could tell you what your dreams were about. She would also dream about people and see things that would occur in the near and distant future. If she were living in today's era, she could very well have had her own TV show! One of the saddest examples of her abilities was her certainty, from her dreams, that she would outlive her three daughters and two husbands; sure enough, she did. This vision took a tremendous toll on her. The only blessing that she realized from it was the fact that the pre-cognition eased the pain of surprise when she received news of each death as they occurred. God works in strange ways that we don't always understand.

Grand-mom developed glaucoma when she was seventy-five years old. As her outer vision grew worse, her inner vision increased. She became a "wizard" of sorts and helped family and friends by acting as a counselor and visionary to all who asked for help.

I never really understood or appreciated the special gifts that women of my family shared. Even though I, too, had these gifts and was utilizing them in my little world, I always thought that my sister, mother, and grandmother were just a little "off." Little did I know that I would be using these same talents in my career and writing and teaching about these gifts in my life as well.

While intuition is often thought to be a feminine energy, it is a talent that we all possess. Whether we use it or not is another story. I understand that the equilibrated energy (male/female) that I recognize in myself has been a great asset to me in my work.

Think of intuition as a muscle. If you use a muscle, you can develop tremendous strength and even perform incredible feats. What you don't use, however, you lose. Intuition is no different. When you focus your attention on your extra-sensory perceptions, you become more and more attuned to the world around you, which can serve you beyond measure. There are several simple actions that can help you discover and refine this beautiful gift,

1) **Acknowledgement:** Recognizing the existence of energy outside of your five senses is paramount to proceeding and developing the skill called intuition. It requires you to step out of the your logical, analytical mind and dive into the sea of creative, imaginative, faith-filled living beyond your five senses. You can only see the seven colors in the light spectrum even though you know scientifically and intellectually that there are more. You can hear a limited number of vibratory tones with your ears, while you know that dogs can perceive different frequencies. Wind exists and touches you although you cannot touch it. You cannot see it, but you can see the effects of it and know it exists. The same thing is true with gravity and with love. There is a famous quote that summarizes this leap of faith into the world of the multi-sensory beings: "For those who believe, no proof is necessary, and for those who disbelieve, no proof is possible." Tuning into your gut feelings, inspired messages, unexplainable signs, and nocturnal visions requires an open acknowledgement that you can, in fact, perceive energies outside of your physical plane and trust the messages as they come.

2) **Rid yourself of doubt:** Push away the logical viewpoint because it will always tell you to disbelieve what you are perceiving.

3) **Expect an answer:** The biggest leap that I or my students make is the one that takes us from the point of *removing the doubt* to actually *anticipating* that an intuitive answer will be forthcoming. It is interesting to examine how an intuitive answer arrives. If you try to force an answer from the universe, there will be a battle between the natural flow of universal energy (where the real answer lies) and the answer created in your logical mind (an answer that is derived from the ego). It is easy to differentiate the naturally flowing answer from the ego-based answer by asking yourself, "Where have my most recent inner thoughts been?" If you have been *imagining* instead of *listening*, it is more than likely that you are not receiving an intuitive message. If your inner most thoughts are along the lines of how impressive the answer will be, I can promise you that you are off course. If you are expecting a certain *outcome* with your answer, such as a miracle cure, or you only want the "wisest" of answers, you will know without doubt that your ego is at work and you are not serving the universe, mankind, or yourself as an instrumental healer—you are serving your ego. I consistently tell my students that all intuitive information comes *through* you not *from* you.

4) **Detach from the outcome:** As an instrument of intuitive readings, you must understand that all human beings have a natural tendency to expect a certain result from their actions. We therefore perceive the ideal outcome *as we would like it to be*—not as God has divinely designed it to be. While positive thinking and expecting a miracle are admirable approaches to addressing your logical mind, it is not

the same procedure as when you ask God a question and *wait* for an answer.

5) **Thankfulness:** As with any gift that we receive, gratitude is the virtue that opens your heart to the grace and light of God. Remember that EGO stands for "etching God out," and that means that you are preventing the Divine Light from penetrating your entire being. The more thoughtful and thankful you are the more light you will receive.

6) **Action steps:** In theory, letting go of the logical, analytical mind and permitting the higher mind to receive intuitive information is easier said than done. Remember, what you are about to experience is really extraordinary, which literally means "out of the ordinary." Like every new challenge, the road to developing your intuitive powers requires practice, patience, and the open-mindedness to learn and trust in the technique. If you follow the first five steps prior to the taking the action steps your learning curve is vastly accelerated.

7) **Maintain an ever-present awareness:** While you are polishing your intuitive powers, be on the lookout for the many "signs" that the universe is sending you all the time. This is what is known as maintaining ever-present awareness. Be aware of the signs when you are driving in your car. A sign can be a roadblock that suddenly makes you deviate from the direction you were going or a house that is positioned on a hill in an unusual manner and captures your attention, or a billboard that has your name on it or the name of someone who you were just thinking about. One day, I was sitting by a river while vacationing in north Lake Tahoe, California. I was contemplating the very important issue of where I was going to open up my *InTuiTouch* Center. Suddenly, a hawk flew

right down the center of the river, going down the river with the current. It grabbed my attention so strongly that I immediately knew it was a sign. All of my recent seminars had been held in Mexico City, and the creation of the seminars had been very easy. I felt that the hawk was giving me a sign to "fly" and "go with the flow." That sign, coupled with other synchronistic events, led me to open my "Oasis of Health and Wisdom" Center in 2005.

Try to discern the hidden meaning behind an event. Constantly ask yourself "what is the meaning behind these events?" The more you maintain this present awareness, the more significant events will show up in your life and the more accurate your interpretations will be. Trusting your intuition is fun and rewarding, and the certainty that you gain from practice creates a whole new avenue of opportunity to follow your natural, divinely designed destiny.

# REAL-LIFE STORY:

*Mary*

In my California clinic one day, I was reading a patient named Mary. It happened to be one of the days when I was really "on." I was very "present" with my patients. I received some information that came through me regarding the cause of Mary's headaches. It concerned her marriage, and Mary just wasn't prepared to acknowledge the truth. Metaphysically, headaches occur as a result of a battle between the higher mind and the lower mind. It is a battle between the heart and the instincts. This struggle causes the minds to create a "mind ache" which we call a headache. Mary's denial caused her to storm out of the office without saying a word to my receptionist. Within about ten minutes, Mary called me from her cell phone and admitted that the truth was just too painful to confront at that moment. The moment that she heard the truth from an outside source (Me) and she knew I had no knowledge of her situation before the reading, she went into a panic. After thinking about it, she was ready to schedule an appointment for a full ***InTuiTouch*** reading and treatment as soon as possible. The underlying cause of her headaches had to do with an unhappy marriage. The struggle was that her heart wanted love and her mind kept saying it was immoral to even think about the idea of divorce. Over the course of the next few months, Mary took the "***The Consciousness of the Heart*** transformational" course as part of her ***InTuiTouch*** treatment and learned how to look at both sides of some of the real issues that she perceived were between her and her husband. By addressing her headaches using the four-dimensional approach (mind, body, emotion, and spirit), Mary learned that behind the headaches was a fear of doing something morally wrong. Behind the issue of doing something morally wrong was a self-esteem issue that had begun in her childhood. Behind the self-esteem issue was a fear that she really could not survive in the world alone. Needless to say, addressing the issues within herself eased the tension between her and her husband, which, in turn, eased her headaches. Mary is grateful for the relief of the headaches that she received from her treatment—but more important—she is grateful to herself for confronting some of her most intimate fears by learning some tools that will help her immensely for the rest of her life.

# CHAPTER SEVEN

## WE ARE HUMAN ENERGETIC VIBRATIONS

# CHAPTER SEVEN

## WE ARE HUMAN ENERGETIC VIBRATIONS

With the assistance of the powerful microscopes that we have today, we have learned quite a lot about our bodies at the "quantum" level. Think about this for a moment—you can break down the human body as a whole into systems like the nervous system, cardiovascular system, etc. These systems can be broken down further into organs like the brain, heart, lungs, and nerves. These can be broken down into tissues, then cells, then organelles within cells, then molecules, atoms, and, ultimately, subatomic particles (protons, neutrons, electrons). When these subatomic particles of our body are broken down even further, what is left is simply vibrational energy or matter that may be even broken down into subcomponents of frequencies of specific light wave particles. Your very essence is pure energy in the form of light wave particles and vibrations. When you understand this, you can understand how vibrations and differences in energy play a huge role in the origin of illness and disease. The lower vibrational tones are denser and have longer wavelengths, which are heavier (there is more matter to them). Higher vibrational tones are shorter and lighter (there is less matter to them).

When your body is operating at a high vibrational tone, it is healthy. We often can see that a person has a healthy "tone" to them. It seems that some things don't seem to bother a healthy person as much because they don't *matter* as much to them. They are lighter and more "in-spired" (which means they have the spirit within) as they move closer to the "Light." (Light is known by other words, such as God, love, truth, health, and balance.) Being in a state of unconditional love is the highest and healthiest state that you can be in.

Illness, on the other hand, reflects a heavier, lower tone that pulls people down because it contains more matter. Things seem to bother or "matter" to them more. The additional weight of matter pulls them down by "gravity" that can lead them faster to the "grave."

Understanding that light levitates and gravity gravitates helps us see how everything, from the microcosm of the body to the macrocosm of the universe, operates on the same principles. We are all drawn toward and depend on light (the source) for our survival. From the most sophisticated organism (mankind) to the single-cell protozoan, we all have light receptors (photoreceptors) that we depend on to bring us the "light" that sustains our lives.

The further away you are from the light (sun, God, source, etc.) the closer you are to the grave (gravity, darkness).

As you delve into energetic anatomy, you will learn about the body's centers of energy (the chakras) and their relationship to health and disease.

These centers of energy direct and control the universal energy that comes from the divine source (God), the Light. These centers influence your physical body (one dimension) and are influenced by your mind (another dimension), as well as your emotions (a third dimension). All three dimensions are incomplete without the fourth dimension, which is your spirit (your connection with the source). There you have it, the four dimensions mind, body, emotions, and spirit. These four dimensions make up the "authentic you," the real person behind the skin that you are in. *InTuiTouch* teaches you to understand that the physical symptoms of the first dimension (body) are inevitably connected to the emotions and the mind (mind-body connections). The mind-body-emotional connections are inevitably connected to the spirit, which I identify here as the "calling," mission, or reason that you are here. When you can identify what your mission or calling is in life and begin to act on it, life takes on a bigger meaning. I believe that there is a divine plan and God doesn't make any mistakes. There is a divine order that is part of your individual plan. Your health challenges are simply a reflection or a signal from your body that you are out of balance.

*An imbalance in your physical body is also a reflection of a proportional imbalance in the other three dimensions.*

You cannot have a change in one part of you that doesn't affect the other parts. You are a whole being. For this reason, the *InTuiTouch* Method is in a class of its own. The principle behind the philosophy of *In Tui Touch* is that any change in one dimension will have proportionate changes in the others. Ultimately, the reason you are here is to learn to love *unconditionally.* This means loving all parts of your life, "in sickness and in health, in richer and poorer, in better and worse"—that is unconditional love that you can understand. Loving your family, yourself, your friends, your finances, your body, and your spirituality are the parts of your life that you are here to learn to love unconditionally. Every one of your life experiences lead to love.

*The more congruent you are in the four dimensions of your "authentic being,"*
*the higher your vibrational level.*
*The higher the vibrational level, the healthier you are.*

## REAL-LIFE STORY

*Eva*

Consider Eva, who had spent six months battling her newly discovered disease, diabetes. Mood swings, depression, and low self-esteem are the result of Eva's extremely low vibratory level. You could even say that Eva was giving off bad "vibes." Life in her world has been very, very dark and very heavy. Weight gain, especially around the abdomen (the third chakra), further supported the conclusion that she was in a very low vibratory level and that her self-esteem was very low. My intuitive reading clearly connected her loss of love with this sad emotional state. It had been nearly a year since her husband of ten years was killed in an automobile accident. It was Eva's perception that all of the "sweetness in her life" was taken away at that instant. She didn't feel that there was really any reason to go on with her own life. As a homemaker and supportive wife, her talent in managing money was utilized in only the smallest ways. While she was fascinated by business concepts of mergers and acquisitions, she had always put her personal desires behind the wishes of her husband for fear of losing him if she pursued her own career. You can guess what had happened. What she feared, she drew near and sure enough, now her husband was gone. Eva felt alone and insecure.

Diabetes Mellitus is a condition that is connected with the ability to control the body's blood sugar. The connection of the loss of the sweetness in her life with her inability to control the "sweetness" in her blood was obvious. In my clinical practice, nearly every diabetic has experienced an emotional trauma that has meant that, in their own perception, the sweetness of life has been ripped away.

Through the treatment on the physical dimension—nutrition, chiropractic adjustments, exercise, and dietary changes—as well as treatment on the mental dimension—connecting the low self-esteem, depression, and despair to the metaphorical sweetness of life—the intuitive reading revealed that Eva's principal energetic losses related to her lack of will to go on. The spiritual dimension involved the identification of her calling, mission, and talent. It was obvious that Eva was bound to become an investment banker. She loved working in this arena, and, after completing courses at a local community college, she was hired by one of the largest investment brokerage firms. From the moment she was hired and began her career, Eva never looked back. The sweetness in life slowly returned, and the love of her husband shifted into a new form of love from family to career. The essence of her love, that she had devoted to her husband, was never destroyed, only transformed—and Eva's energy simply shifted from one form to another.

Incidentally, with her *InTuiTouch* treatments, she began a new dietary course, started to exercise, and found an entirely new outlook on life. After six months of treatment and another six months of self-care, Eva's blood tests revealed normal blood sugar levels and absolutely no trace of diabetes.

The *InTuiTouch* Method of healing occurs in all four dimensions of health: the spiritual, the mental, the physical, and the emotional.

# CHAPTER EIGHT

## THE SEVEN BASIC INGREDIENTS FOUND IN THE HEALING SOUP

# CHAPTER EIGHT

## THE SEVEN BASIC INGREDIENTS FOUND IN THE HEALING SOUP

In order to understand how InTuiTouch Method can help you or someone you love improve in health, you need a clear understanding of this thing called "healing." First of all . . . what do we mean by "healing"? It is my opinion that no one is ever healed! While this statement may disturb you at first, let me explain. No one is ever healed: they are healing! It is an ongoing process in the three dimensions of mind, body, and spirit. Your body, even at the cellular level, is not the same as it was even yesterday! That broken leg eight years ago actually caused a production of bone cells—it caused the leg to grow new bone. Those new cells have since been replaced with new cells, and the process repeats every second of our lives. So did we heal or are we still healing? That high fever you had as a result of an infection was designed to kill off the virus or bacteria; by elevating your body temperature, it created an environment that the invader couldn't live in. Do you stay at that high temperature level indefinitely? Of course not! Soon the temperature returns to normal. Does it then stay at that temperature? No! Sometimes it your temperature drops, also. Your body is constantly going through a process of adjustment to your environment. Your mind and your spirit are constantly going through adjustments as well. This is all part of the design, to help you maintain a more congruent, balanced way of living. All of your experiences, yes, even those injuries and illnesses, have a reason for occurring in your life. They are there by divine design, helping you learn certain lessons that you need to learn. That sprained ankle is trying to slow you down and make you take a look at where you are going. That rash on your arm is there to remind you to love that person who is "getting under your skin." That auto-immune disease is reminding you to stop attacking your self-worth.

The healing ingredients begin with what I call the "attitudinal spices of life's soup."

**INGREDIENT #1: OPEN MIND**—Healing begins in the mind and your attitude; how you perceive yourself in relation to the world plays an important role. You must get over the idea that you are a "victim" of an accident or bad luck or genetics. Open your mind to the idea that you are not being cosmically punished; stop thinking that somehow God made a mistake and gave you the raw end of the deal. If you or someone you know has a health challenge that conventional medicine has told you has limited options, open your mind to the truth that medicine is a *science* and is founded on principles that *change* according to man's continued research and further understanding. A medical opinion is never an absolute, and the rules keep changing. Remember, it wasn't that long ago that the cold, sterile environment of the hospital operating room was considered the "scientific" way to birth healthy babies. Now the hospitals are competing for your birthing business by offering their newest versions of "natural birth." Healing is an *art* established by Mother Nature and founded on universal laws that *never* change and have withstood the test of time. Keep an open mind that there are other paths to healing.

**INGREDIENT #2: PROPER FUEL**—The expression *"You are what you eat"* is not completely accurate. The truth is you are what you *ate*, and you will become what you *eat* now. There is no coincidence that the leading causes of death in the United States (heart disease, cancer, and diabetes) are directly or indirectly related to diet. Think about it. If your body is designed to use a certain fuel to function at its maximum capacity, and you constantly feed it lower-grade fuel (junk food, stimulants, sugar, etc.) it is only a matter of time before the body will give off signals that something is wrong. Eventually it will break down. In health care, we call those signals "symptoms." What you put into your system does make a difference. Try to honor your physical temple as much as possible. I know you're wondering whether that means that you have to become a health-food fanatic or vegetarian. No!

You don't *have* to do anything. It is all a matter of *choice*. It's an attitude. If you have any health challenges right now, take a look at what your diet has been and search for the connection. Most nutritional experts agree that lighter the meal (vegetables and fruits), the easier it is on your digestion. The fresher the food that you buy, the better it is for you. It seems that the more "alive" the food is, the more life it feeds you with. Meats (dead animals) have a tendency to weigh you down, and they command an enormous amount of energy just to digest. Remember this quote: "It's not what you do between *Christmas and New Year's* that makes a difference in your health, it's what you do between *New Year's and Christmas* that makes the difference." The trick to having a balance in this area is to eat with a sense of moderation, simplicity, and rhythm. Whether you have a current health challenge or you are focused more on wellness and prevention, eating the proper foods is essential in keeping the physical dimension of that triad in the best operating order. What about vitamins and other supplements? In my book *The Seven Myths of Nutrition*, I talk about the fact that our farming methods of chemicals, pesticides, and other contaminating factors may increase the production of foods in numbers, but it decreases the nutritional quality of that same food. Research has shown that the farmlands in which we grow our foods are grossly depleted of the necessary minerals and nutrients. We, therefore, are naturally depleted in the more than a hundred and twenty essential vitamins, minerals, amino acids, and enzymes that our bodies need to function at 100 percent capacity. It is no wonder that we have developed so many devastating illnesses, particularly the chronic degenerative types. In this day and age, nutritional supplementation is not a *Luxury*, it is a *Necessity*—especially if you have an already existing health challenge.

**INGREDIENT #3: DESIRE**—Zig Ziglar, one of the most popular motivational public speakers of all time, said, "Your attitude and your aptitude affect your altitude." When it comes to healing, the desire is an essential ingredient of this soup. Remember, too, that *there is no doctor who can do this for you*. Desire is an "inside job." The way to raise your motivation or desire to heal is by *linking* the *desire to heal* with your highest values in your life. Are your children important in your life? Do you have a mission or calling in life that you must fulfill? What about marriage? What would happen to you those values if you were not able to continue because of health challenges? Whatever you value most highly, try linking your health to those things.

**INGREDIENT #4: HUMILITY**—I stated earlier that EGO stands for "*E*tching *G*od *O*ut." If you etch God out of the center of the universe and you let your ego get too high, the natural homeostatic mechanism will be at work to humble you again. Anyone who has had a major illness of any type will tell you. Humility teaches you to "Let go and let God," so that you may go with the flow of life and not against it. In order to allow healing to occur, you must go with the flow of nature (God)—not against it. Humility teaches you to cast no judgment on others for your challenge and helps you eliminate the "Poor me" syndrome. Humility does not have to be practiced only when you are in a crisis situation. It is, however, a conscious decision. Humility in its best form will lead you to the next ingredient.

**INGREDIENT #5: GRATITUDE**—When you look at your life or even your health and compare it to others', you will perceive some who are suffering less and some who are suffering more. It is then that you realize that your life's challenges are relative, and even you have been blessed. Counting your blessings is the best way to open the door to gratitude and connect to your higher power. Gratitude elevates your vibrational tone, which allows all of the cells in all of your systems to resonate at their highest and healthiest levels. It also sets the stage for healing to begin by bringing you closer to God, closer to the Light, and closer to the healing source.

**INGREDIENT #6: OPEN HEART**—Humility and gratitude are the keys to allowing the heart to open. It is only through an open heart do you align yourself with that highest power, God, which is the source of all healingand everything else as well. When your heart is open, you are in a state of love. You are not holding onto any judgment about others or self. You are showing compassion for others as well as yourself. It is a state of grace, which in Spanish is called *gracias* or thank you. Love is key ingredient to all healing, and only through an open heart can you allow love to enter and permeate your being. It is only through love that you will are able to align the body, mind, emotions, and spirit.

**INGREDIENT #7: TRUST YOUR HEALTH CARE PROFESSIONAL**—Even though we understand that healing comes from within through the divine innate source of wisdom that runs our body, the relationship with your doctor or other health professional is imperative. Trust that he or she has your best interest at heart and that he or she is only the instrument from which the science of healing can be integrated with the art of healing. Your health professional should be humble enough to recognize that the healing power comes *through* him or her and not *from* him or her. An interesting fact was printed in the *Journal* *of American Medical Association* (JAMA) regarding patients who left one doctor for another. It was not because the doctor didn't have experience and it was not because the doctor's fees were too high or the doctor was not covered by their insurance. The number-one reason that a patient looks for another doctor is *doctor apathy*. The patients just didn't feel that the doctor was interested in their care, or treated them as one human being should treat another. Look for a doctor or health professional who you feel you can trust and who is looking out for your best interests at all times.

These are the seven ingredients that make up the healing soup. If you add them to your meal called "life" when the need to heal arises, you will experience incredible healing changes in your body. Remember—we are here on the planet to learn about love, life, health, and living in all imaginable ways.

Remember—we are here to learn about life in all of its forms.

# CHAPTER NINE

## HOW OUR "BIOGRAPHY" AFFECTS OUR "BIOLOGY"

# CHAPTER NINE

## HOW OUR "BIOGRAPHY" AFFECTS OUR "BIOLOGY"

Carolyn Myss, PhD and author of Anatomy of an Illness, coined this phrase, "how our biography affects our biology," to explain how our emotional experiences and perceptions influence the physical functioning of our bodies. There is a great deal of wisdom in this observation. Can you see where this applies to your own life? Any emotional memories that you may perceive as wounds or baggage affects your body physically. I learned in medical training that "muscles are the seats of emotion." You may harbor deeply buried emotions in certain muscles, which might then appear to be chronically tight with very little elasticity, or have great sensitivity (hypersensitivity) when lightly pressed. Each muscle group in the body belongs to the "body map" explained in the InTuiTouch Method. An example is that the tightening of the trapezius muscle relates to burdens and guilt. Stomach muscle knots relate to self-esteem issues relating to anger and resentment from the past. Even the heart, which is also a muscle, will elicit aches and pains if the perceived past experience was heartache. No one escapes the tragedies of life. We are not supposed to. You cannot avoid them, even if you try. Remember: What you fear, you draw near, and what you resist will persist. While you cannot avoid the tragedies in your life, you can alter your perception of those tragedies. You can look at each event in your life that you have perceived as bad and find the benefits of those events and see how they have served you today. In his audiocassette tapes, The Secrets to Enlightenment, Dr. Wayne Dyer states, "Life gives exams . . . If you don't pass, you repeat the course."

My parents divorced when I had just graduated from high school. Their lack of communication led them quietly to the destruction of the love that had brought them together in the first place. Because of what I perceived as an ugly experience, I repeatedly stated, "When I marry, I will never get divorced!" As much as feared divorce, I repeated the same challenge—not communicating my needs as clearly as I could have—out of a fear of rejection. My biological challenge was self-esteem and the fear of abandonment. My biography was definitely affecting my biology. The more I *im*pressed my feelings, the less I *ex*pressed them outwardly in the form of communication with my wife—who became my ex-wife. Stomach pains, irregular heartbeats, and a constant need to "clear my throat" were the symptoms that ran my life at the time. Instead of communicating my needs to my wife, I began to communicate them to another woman whose infatuation with me gave me a false sense of high self-esteem. After ten years of marriage, I was drawn into a three month pseudo-affair, from which my ex-wife never recovered. Ultimately, after twenty-one years, I attracted exactly what I feared into my life. Today, with the help of the ***InTuiTouch*** Method and the ***Consciousness of the Heart*** transformational course, I have found the way to "pass the exam" that my life is giving me. One of the most powerful lessons that I learned is the importance of a clear and open dialogue about any challenges that I encounter with a partner. I want my partner to be my best friend. She should be my blessing in my crisis. I can now pat myself on the back for understanding this, and benefit from all of work that I have done to get out of that class in my life school—understanding unconditional love.

# CHAPTER TEN

## AN OVERVIEW OF ENERGY, AURAS AND THE CHAKRAS

# CHAPTER TEN

## AN OVERVIEW OF ENERGY, AURAS, AND THE CHAKRAS

*There is more to you than just the skin that you are in.*

I n addition to the physical aspect of your being human, there exists a subtle, yet powerful, energetic system that is located within and around the physical form that we call the body. This system consists of seven subtle layers of energetic fields known as auras, and seven principal vortices of energy known as the chakras.

The auras can best be described as the manifested expressions of the human spirit. They are all connected with each other within the same person, but they also extend out to touch others. Many people can see these fields in people. Auras are commonly seen by children. Children describe their experience as *"seeing* colors around people," and this is known as clairvoyance or extra-sensory perception (ESP). ESP can also be *felt* by some people. This type is known as clairsentient. While it is true that some adults have a heightened awareness of these fields, everyone has the ability to develop and refine this extra sense. To demonstrate your ability to experience this energetic field, try this:

## ESP EXERCISE

Sit facing a partner, and extend both hands out to lightly touch palms. After a few short minutes, you will be able to feel an energetic wave between you. Now, while standing, separate yourself by approximately three feet, and, with arms extended toward each other, begin to feel the energetic vibrations as you come closer and closer together. Typical sensations reported are sweaty palms or tingling. These are indications of the energetic waves that interconnect when we approach another's auric field. We sometimes call this "stepping into someone else's space."

The word *chakra* comes from the Sanskrit word meaning "circle." Children and adults who have the ability to see chakras describe them as colors moving in a circular fashion over specific parts of the body. The chakras serve as receiving and broadcasting stations that literally connect the physical world with the spiritual world. The chakras are the gateways that link the emotional and spiritual energies to our physical bodies. These emotions and attitudes run through the chakras and are distributed to every system, organ, tissue, and cell in the body. Acknowledging the chakras is of prime importance in benefiting from the *InTuiTouch* Method and philosophy of healing. When you can link the emotions and corresponding mental attitudes to the spiritual "inner calling," the physical bodily expression of those energetic changes can not only be understood, but also *experienced;* thus, healing occurs. Each of the seven chakras has a vibratory connection to one of the seven colors of the rainbow, as well as one of the corresponding seven glandular connections of our endocrine system. Different levels of consciousness are connected to specific chakras as well. Even the foods you eat and their corresponding colors will affect different chakras. There are even specific-colored gems and minerals that act as catalysts to draw in more energy to specific chakras.

*By understanding the chakras and their role in linking the mind, body, spirit, and emotion, we can better understand how any health challenge—regardless of the diagnostic name attached to it—is only a physical expression of a deeper, more integrated loss of energy.*

The more you know of this connection, the more control of your own health you will have. In a sense, you could say "to know yourself is to heal yourself."

The seven principal chakras are located in the aura level known as the "etheric body." It is the closest auric layer to the physical body.

# CHAPTER ELEVEN

## THE CHAKRA SYSTEM

# CHAPTER ELEVEN

# THE CHAKRA SYSTEM

The seven principal chakras, located in the aura level known as the etheric body, each have unique qualities, characteristics, and functions. These interact with the other chakras, creating a network for the life force energy that connects our mind, body, emotional, and spiritual dimensions.

**Chakra 1:** The root chakra is located at the base of the spine near the tailbone. It draws vital force energy from the earth and connects your very basic survival needs (food, clothing, and shelter) to the higher, more spiritual needs (inspiration, gratitude, unconditional love, etc.). This chakra ties you to the needs of belonging to a group, such as family, friends, and community. A healthy first chakra will allow you to draw from the deepest sources of vital energy. In this way, a healthy first chakra can enable you to utilize the power of hands-on healing for self and others. You may also use this energy to be the most creative of persons in any field; this vital source of energy allows you to magnetize people, places, and events in life to create anything imaginable. Energy is lost when you feel your loss of connection to mother earth or the most basic needs of survival. The "victims" of the world have lost energy in this first chakra. The "motherly instincts" support the energy in this chakra.

Losing energy in this chakra produces physical symptoms involving the lower spine and pelvis, the colon, legs, kidneys, and adrenal glands. The emotional symptoms of insecurity, violence, fear, and anger are related to the base chakra.

Chakra 2: The sacral chakra is found just below the navel. It controls our vital needs for money, sex, and power. A healthy second chakra would find you with a sense of well-being and feeling of having enough of everything in your life. You have learned to give as well as receive, and you function well with respect to passion, desire, sexuality, and pleasure. Physical symptoms relate to the ovaries, testicles, bladder, prostate, spleen, and genitals.

Losing energy in this chakra produces emotional imbalances displayed as over—or under-indulgence in food or sex, impotence, jealousy, and the need to control. The "martyrs" of the world have lost energy in this chakra and the "emperors" of the world have mastered this chakra.

**Chakra 3:** The solar plexus chakra is located just above the navel in the region of the stomach and abdomen. It serves the function of digestion of both the foods that you eat and the emotions that you consume. This is the center of your self-worth, self-esteem, and personal power. Unhealthy third chakras will stem from ego-driven imbalances and will result in an intense storage of deep-seated emotions.

Stomach problems, such as ulcers and indigestion, are clear signs of energetic loss. The pancreas, liver, gall bladder, and adrenal glands are all connected to the third chakra. The "less than or greater than you" personality, which is an unavoidable pendulous swing that everyone experiences as "low self-esteem" or "self-righteousness", will reflect strong energetic losses of this center. The "Sensei Warrior" personality of wisdom, patience, and humility reflects a balanced third chakra.

**Chakra 4:** The heart chakra is located in the chest area, centered near the sternum or breastbone. It is clearly the most important chakra of your energetic being, as it represents unconditional love of self, others, and the universe itself. It is the center for forgiveness and compassion. Just as the heart serves as the most important organ of your body, the heart chakra serves as the connection between your physical instinctual chakras (one through three) and your spiritual essence chakras (five, six, and seven). You often express your deepest meanings in life and love through your heartfelt experiences. The biblical expression states that "in order to give, one must be able to receive as well, and in order to receive, one must give." You must first learn to love yourself unconditionally before you can spread that same love to the world. Many people with low self-esteem will learn to give but not receive. Others, who have acquired the ability to close their emotional hearts in order to not feel the pain of heartbreak, develop congestion from lack of expression, which results in conditions like congestive heart failure and heart attacks. The actresses and actors of the world portray false images of love on a very superficial and safe level, while the true lover who is willing to risk it all for love gets to experience the full joy and wonderment of unconditional love.

**Chakra 5:** The throat chakra is the vehicle by which you are able to voice your truth, demonstrate your self-expression, exert your will power, and, ultimately, communicate your feelings and thoughts to others. Just as the physical throat is left unprotected simply by its anatomical location, the throat chakra is the most easily damaged by negative experiences and your own perceptions of events that have occurred. Whenever a child is scolded, the child instinctively lowers his or her chin and protects the throat chakra. As adults, we mask what we really want to say by covering our mouths with our hands or even grabbing our throats as we speak. Some people repress their inner feelings by masking the throat chakra with excessive eating, smoking, alcohol consumption, or drugs.

Health challenges in the throat, thyroid, teeth, lips, and bronchial tubes are typical signs of an unhealthy fifth chakra. A person with a healthy throat chakra is the individual who speaks up for what he or she wants with clarity and integrity.

**Chakra 6:** The brow chakra is located between the eyes in the area commonly known as the "third eye." It is the center for thinking, analysis, imagination, and intuitive powers. The development of intuition and psychic power resides in this chakra. We all strive for balance of the right (creative) and left (analytical) brains so that we are able to optimize our connection with the earthly experiences as well as the heavenly or spiritual ones.

To really reach the heights of well-being and satisfaction, you need to have a healthy sixth chakra. The more developed you are in this area, the closer to fully actualized you will be. Energy losses in the sixth chakra usually are the result of an over-intellectualized side of oneself. A brain filled with data is ego-based and therefore unappreciative of life as it is occurs. An underdeveloped left brain and over-utilized right brain will cause confusion and doubt with very little planning, resulting in a day-to-day survival mode of living that causes overall growth to stagnate. People who lose energy in the sixth chakra can develop neurological conditions, such as epilepsy, Alzheimer's disease, multiple sclerosis, and others. A healthy sixth chakra will lead to opening the seventh chakra, which will clear your connection with the universal intelligence and spirit.

**Chakra 7:** The crown chakra can be viewed as the "crown jewel" of all of the chakras. It is the essence of our spirituality. It is the "Christ consciousness" and the connection to the infinite source of light and love.

Energetic losses at the crown level are usually the result of an ego that has gone out of control or arrogance run amok. The person who believes that he or she is fully responsible for all of life's blessings and who fails to acknowledge the existence of spirituality or a higher guiding force has a weak seventh chakra. The classic symptoms of energetic losses in the seventh chakra are depression and lack of emotion. When it is apparent that all of life's goals have been achieved, and a person takes personal credit without the acknowledgement of a higher power, an emotional "ceiling" is reached in the journey toward self-fulfillment; depression sets in. Gratitude and humility are the only tools that can cure this type of infirmity. The well-balanced seventh chakra, on the other hand, is evident in the person who has mastered his spirituality, recognizing a oneness with the universe and acknowledging that nothing is missing. Therefore, nothing is needed. This person has mastered the law of magnetism and attraction and has conquered the human weaknesses of judgment and criticism. Universal energy flows through this person and, even though the lives others are changed for the better in all that he or she touches, no personal credit is accepted.

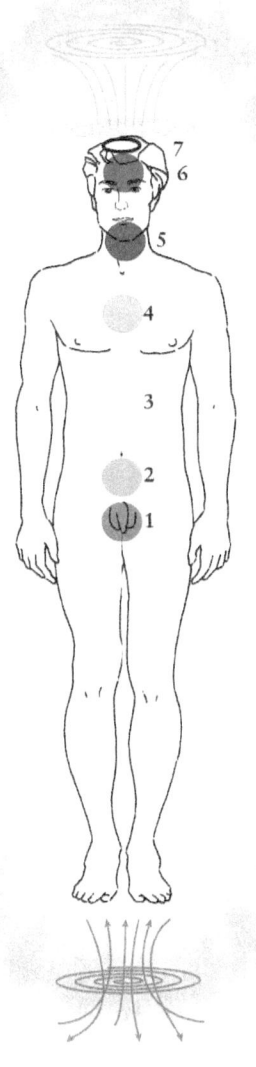

# CHAPTER TWELVE

## LOSING ENERGY COSTS HEALTH

.

# CHAPTER TWELVE

## LOSING ENERGY COSTS HEALTH

When you function with less than 100 percent energy, health is compromised. Let's look at seven facts that support the philosophy of the InTuiTouch Method:

**1: When you break down a human body to its essence, you see that we are simply energy.**

**2: *Emotions* (energy in motion) swing from positive to negative.**

**3: Past guilt and future fear cause you to lose energy and keep you out of the "present."**

**4: Lost energy *disconnects* mind, body, emotion, and spirit and works against integration.**

**5: *Dis*-connection causes *im*-balance and is the root of all *dis*-ease.**

**6: Lost energy affects your physical body at the cellular level.**

**7: All glands, organs, and systems function poorly on reduced energy.**

In the previous chapter, you learned how we receive 100 percent energy from the universal source (God) in the present. As present time is diverted to past guilt and future fears (in equal proportion) based on our perceptions of our experiences, we lose energy in the present. In the physical form, our cells operate on the vital energy provided by the universe. As this energy is diverted into past and future imaginings, our cells operate on *less than optimum* amounts. Any amount less than 75 percent of the optimum amount results in the cells of our body sending us signals of energy deficiency. We call these signals "symptoms." Persistent deficiencies of vital energy to our cells results in what we call disease or *dis*-ease and can ultimately result in auto-destruction of our organism.

# REGAINING ENERGY RESTORES THE HEALING PROCESS

*Physical energy is restored by proper treatment, rest, diet, and exercise.*

*Emotional energy* is restored by acknowledging the fact that emotions are swings in one's perception. According to Dr. Demartini's Quantum Collapse Theory, emotions are exaggerations and minimizations of the truth. The truth is a dualistic, balanced perspective allowing one to see both sides of any event that occurs in life. In a person's *reality*, the a person, place, thing, or event is perceived as either positively charged (happy, good, inspiring, etc.) or negatively charged (sad, bad, despairing).

A person's *actuality* sees an equal and balanced number of positive and negative aspects of the person, place, thing, or event. Any time you perceive more negatives than positives or more positives than negatives, your mind is taken into the past guilt or future fear, which prevents you from being in the "present." This emotional swing can, in essence, run your life for a brief moment in time—or many years.

# REAL-LIFE STORY:

*New York's Twin Towers*

If you see only negative associations with the events of 9/11 in New York, because of the nearly three thousand deaths, you miss out on the magnificence of how our country became united, how donations from people's hearts poured in, how security in airports improved, and how countries united around the globe. You would have missed out on the fact that the "blinders" were removed from US citizens' eyes about the vulnerability of not just the United States but any country. The *InTuiTouch* Method focuses on the philosophy that true healing of emotional and mental energy loss begins when you seek a balanced perception called (truth) in any event that occurs in your life.

This is known as your *Actuality* instead of your *Reality*. When you can find the complementary opposite charge (opposite emotion) in any person, place, or event where the truth is being masked by your lopsided perception, energy is shifted and restored at the mental, emotional, and even spiritual dimension.

## REVIEWING THE CHAKRAS

(See the following illustrated chart) Each of the auric layers maintains a different vibrational tone. The highest vibrational tone is the divine body and the lowest is the physical body.

# The Layers of the Aura

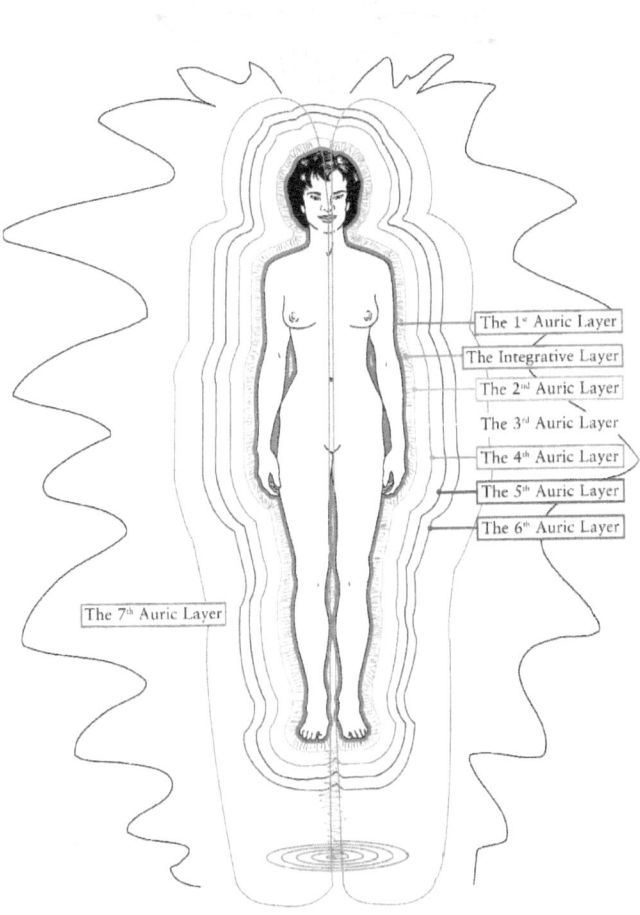

The 1st Auric Layer
The Integrative Layer
The 2nd Auric Layer
The 3rd Auric Layer
The 4th Auric Layer
The 5th Auric Layer
The 6th Auric Layer

The 7th Auric Layer

# INTUITOUCH

# CHAPTER THIRTEEN

## REAL LIFE STORIES

# CHAPTER THIRTEEN

# REAL-LIFE STORIES

## "IT'S ONLY AN ANKLE SPRAIN"
*Sue*

When Sue came into my office, hobbling, with a badly sprained ankle and lower-back pain, she knew that her injury was more involved than what her medical doctor had told her. His examination and X-rays on the day of the injury revealed only a ligament injury that should have responded after two weeks—unless there was more going on with her "injury" than she was willing to see.

The metaphysical connection of the left ankle was clear. Ligaments are our *connective* tissue. Swelling and *inflammation* led me to ask the question, "Who are you *connected* to that you are *inflamed* about?" The mind-body connection of the left ankle has to do with moving forward in life, relating to the heart and emotions. Energetically speaking, Sue's lower-back injury involved the second chakra, the sacral chakra, which is connected to her one-on-one interactions with her husband. Her body (the physical expression of who she is) was telling her to stop and take a look at where she was going (because the ankles give us "flexibility in movement") in relationship to love and her heart. She was inflamed with her husband and herself because they had let their professional lives strain the energy of their marriage.

As it turned out, her district managerial position demanded a lot of time away from home and, of course, away from her husband. While her husband, Douglas, can be a most considerate man who understands that nature of her career, even he admitted that their relationship had been suffering. This injury was literally slowing Sue down, which gave her a real chance to stop and take a look at her life. She was afraid to cut down on her travel time and jeopardize her work, but she knew that if she valued her relationship enough, something would have to be addressed eventually. It was easy for Sue to see the connection. It hit her like a ton of bricks. Sue's homework was to talk to her husband about plans for quality time with him while listening to his needs, desires, and wants as well. To her amazement, within the next twenty-four hours, her swelling disappeared. The lower-back pain disappeared over the next two days and Sue was back to work in three days.

Was Sue a victim of an accident to her ankle and lower back? Was it coincidence? It was nothing of the sort. By divine design, Sue had created an opportunity to get her life back into balance, something that she wanted to do but was afraid to address. Her fears of "not having enough" income and security were, in a sense, running her life, and it was time for her to look at these challenges.

# THE ISLE OF SOMEDAY

*Joanne*

Joanne continued with her chiropractic care even though her symptoms continued. Her neck and shoulder pains felt better with the treatment—but only temporarily. Her headaches persisted, and she learned to live with those just as she had learned to accept her irregular menstruation. While her adjustments were helping her in the physical dimension, the mind-body-spirit connection was overlooked again. Headaches are actually "mind aches" and point us toward the internal struggle between the lower-minded survival instincts and the higher-minded "inner calling" of our spirit. Joanne was battling a career change. Her talents were in life coaching but she was afraid to step into that "unknown" and take a chance. Her financial responsibilities were perceived as the weight of the world on her shoulders, which connected intricately with her shoulder tension. The fear of actually facing the truth of her calling manifested itself in the form of neck pain. She stated, "It only hurts when I turn to the right." The right side of our metaphysical body map relates specifically to business and financial decision making. The left side of this emotional map relates to matters of the heart. Joanne was tied deeply to her present job and was living on "Someday Isle." "Someday, I'll make the change." "Someday, I'll get rid of all of these bills so that I can do what I love to do." Her inner voice was very clear and wasn't going away. In fact, the increase in symptoms was directly related to the increase in acknowledgement of her "calling."

Intuitively, I knew that Joanne needed to be congruent with what her heart and what her body were telling her. She was losing energy from her sixth chakra (applying her wisdom, knowledge, and talents) and from her fifth chakra (speaking her truth and applying her will power), and from her second chakra, the sacral chakra (the challenges of money, sex, and power). She left the very first treatment session with a sense of relief just from acknowledging what she already knew. Within two weeks, Joanne had enrolled in a life coaching course, and she was determined to phase out of her current job and phase into her calling. Is it any surprise that Joanne began to experience dramatic improvement while continuing her chiropractic care? Within eight weeks, she was completely symptom free, eternally thankful, and loving life to the fullest.

## GONE IS THE SWEETNESS OF LIFE

*Bob*

The sudden passing of his wife was more significant than Bob ever imagined. He didn't even connect the fatigue, weight gain, and mood swings over the next few months to a diagnosis of diabetes until he passed out from insulin shock on the eve of his birthday. Within two years, Bob's eyesight was deteriorating, as was his overall health. Through the *InTuiTouch* Method, I was able to read which of the chakras were losing energy, and I identified Bob's diminished will to live. He had literally lost the sweetness of life, which correlated chronologically very clearly with the death of his wife. Literally, diabetes is a blood sugar control problem involving the pancreas and its ability to produce adequate amounts of insulin. Physically, Bob was placed on a very strict diet and a serious vitamin/mineral supplementation program designed to help his body help itself. Energetically, Bob was losing significant amounts of energy in his heart (fourth) and solar plexus (third) chakras. Through the process of the Transformation technique, Bob was able to see the benefits and the drawbacks of the events over the past two years. By identifying his calling in life, Bob began to turn his life around. With his diet under control, his energy returned, his weight dropped, and he even began dating again. You could even say that the "sweetness of life" had returned.

The focus of these stories is to demonstrate the value of addressing the "authentic you" in the four dimensions of mind, body, emotion, and spirit. In each and every case, there was a necessity to acknowledge these as integrative and inseparable. When this is accomplished, true healing begins.

# CHAPTER FOURTEEN

## OUR AURA ENERGY FIELD

# CHAPTER FOURTEEN

## OUR AURA ENERGY FIELD

Make no mistake, our bodies are divinely designed to mirror of the rest of the universe. We are energy, pure and simple. If you think about it, why would we be structured any differently than the universe, where a harmonious order already exists? The human aura system is a perfect miniature duplication of the larger-scale universal energy system. It is the nature of our very existence. As Victor Hugo so eloquently stated in 1862, "Where the telescope ends, the microscope begins . . . but which of the two has the grander view?" Auric energy is believed to begin from the energetic center of each of our individual beings: the solar plexus region. This is the source that powers the four dimensions of each authentic individual, the physical, mental, emotional, and spiritual It is comforting to know that we are always connected to the cosmic or universal energies (God), and, in fact, we are one with the universe while we are separate energetic forces. This can be compared with the idea that the air in one room is separate from the rest of the house but is still part of the larger body of air that permeates the entire house, city, country, and world. We can extend our energetic aura (as in the case of public speaking or acting), and we can withdraw our aura (as we would do in a crowded room or an uncomfortable situation).

As a constantly moving energetic field, the human aura incorporates seven bodies or expressions of the human spirit, each vibrating at a different rate. Each expression contains one of the seven corresponding colors as well. While the color, size, and intensity will change, depending on life circumstances, there remains a stable, predictable auric structure and personality type that remains as your "energetic blueprint."
Many children, and even some adults, see auras as colors or sometimes see auric shadows around people. Some people can only sense this energetic field. We all have the ability to perceive this field, but only a few acknowledge and refine this gift.

The profound exploration into the energetic origins of our being has been vastly enhanced through specialized photography. Initially this occurred using the Kirlian photograph, which captured the energetic layer around plants, animals, and ultimately humans, but now, several generations later, there exists the *auric camera*, a highly sophisticated camera that captures the colorful auric layers of each individual. Highly trained intuitive interpreters can now read your aura and offer fascinating information regarding your past, present, and future. The **In Tui Touch** Method utilizes the auric camera as part of the metaphysical diagnosis that each individual receives during treatment. Due to its subtle energetic detection, the camera serves as a guiding light, providing valuable information regarding all four dimensions of the "authentic you."

## NEGATIVE INFLUENCES ON THE AURA

Just as stress, anger, and hostility create eventual breakdowns in our physical body, these emotions create wear and tear on our energetic systems as well. Similarly, negative memory patterns and experiences that lead to poor self-esteem, lowered personal value, introversion, and antisocial behavior all have their corresponding drain on our energetic body as well.

## POSITIVE INFLUENCES ON THE AURA

The most influential power in the universe is *unconditional love*. The more loving and caring an individual is to himself and others, the more radiant his aura becomes. Other attributes that contribute to one's healthy aura include a sense of well-being, balance in one's life, and the recognition that one is connected universally to the source. This is demonstrated through acts of humility and gratitude toward self and others.

**Soulmates' auras, matching energy**

**Bright ideas**
**Red left: shows vitality**
**coming into his aura**

**Violet/pink: unconditional love**

**Indigo blue: old soul
much wisdom to impart**

**Multifaceted/creative
Green: shows teaching power**

# CHAPTER FIFTEEN

## INTUITOUCH IN ACTION

# CHAPTER FIFTEEN

## INTUITOUCH IN ACTION

### THE MIDAS TOUCH
*Tommy*

Tommy has known for a long time that when he put his hands on people, they felt better. It was like a hobby for him. Whenever his family or friends would gather, inevitably someone would ask Tommy to work his magic. More often than not, the results were amazing. Even at work, while he was on sales calls, customers would ask Tommy to help them—and of course, he would oblige. During a session of InTuiTouch, Tommy learned through his auric reading that he did, in fact, possess an enormous amount of "red energy" in his aura. Red energy translates into a specific connection to the Vital Force energy that is found in "hands of healing." This confirmation via his auric photograph inspired Tommy to pursue his dream of entering into the healing profession. He now recognizes what God-given talent he has. Because he is more congruent with the cosmic universal energy, he is magnetizing circumstances and people into his life that he previously only dreamed of. His life has changed dramatically, not only in his career change; he is now married and has a new baby.

## ARTHRITIS? SHE CALLS IT ARTH-WRONGIS!

*Rosa*

Rosa was diagnosed with rheumatoid arthritis and told that it would progress to the point of immobility over the years. She was offered no hope offered, merely an increase in the dosage of her anti-inflammatory medication and pain pills. None of the treatments seemed to be working. Her surgically replaced knee joint did not help her with ambulation or pain. During her InTuiTouch treatment session, Rosa's aura was photographed. She was able to see the "guardian angel" or spirit guide that was so obviously positioned as a pure white light right by her side in her aura. "I always knew I was being protected by my guardian angel," she boldly stated. "This illness is going to get better, and I don't care what the medical doctors say!" After twelve sessions of *In Tui Touch*, Rosa's arthritis was virtually nonexistent.

While the information obtained from the auric photograph is not and cannot be used to diagnose or treat any medical conditions, it serves as an invaluable tool in the metaphysical diagnosis of energy in the four dimensions of the authentic being. Healing at this level transcends the mainstream approach and can be a valuable adjunct to any medical care as well.

# CHAPTER SIXTEEN

## HOW MUCH OF A ROLE DOES ENERGY PLAY?

# CHAPTER SIXTEEN

# HOW MUCH OF A ROLE DOES ENERGY PLAY?

Following the reasoning that we are four dimensional beings and that each of the dimensions is of equal proportion and value to our authentic being, there are some interesting questions raised as to whether the cause of an illness can be rooted in any one of the four dimensions initially and then "spread" to the other three dimensions. InTuiTouch maintains that all of the dimensions are affected simultaneously. We know that, at the energetic level of our being, our spiritual essence exists in a timeless, space-less, mass-less, and charge-less state.

Therefore, changes that we perceive in our physical body as "illness" truly affect the four dimensions simultaneously.

## REAL-LIFE STORY

### "HE DIDN'T GET THE LESSON"

*Richard*

Richard was playing racquetball with his usual Tuesday lunchtime partner. After showering and a little lunch, he returned to his car to make a call. Suddenly, he experienced a series of sharp chest pains that led to a heart attack. Richard never made it back to work that day; he died right there on the spot.

A fierce competitor and also a fearfully jealous man, Richard was constantly comparing his social status with others'. Although his business was sinking, his pride forced him to pour what little he had left in savings into advertising, just to keep up with the other office-supply stores in the community. Owning his store was not his real passion, which was obvious in the way that he always looked for ways not to be at work. His real passion was growing roses. Next to the love he had for his family, nothing ever came close to the love he had for raising and showing off his different hybrids of roses. The contributions that he had made to the Rose

Society left an indelible mark on the way roses were grown and presented. Richard recognized his talents but ignored the calling. He was locked into the fear of the future, always worrying about not being able to provide enough income from a business created from his passion for flowers. He struggled on with his office-supply company despite the many warnings and obvious signs of the times. Franchised mega-office-supply stores had already reached his tiny town and made a serious dent in his profits.

His ego and reputation for "not failing" created an ever-mounting level of stress, which also contributed to the heart attack. A "calling is a calling," and you can ignore a calling for only so much time. We begin to receive signs from the source of our gift and talent. At first, we get little nudges or small signs, letting us make the decision to be true to our calling and listen to that inner voice called intuition.

Dr. Demartini in his book *The Breakthrough Experience,* states that "a genius is one who listens to his soul and obeys." Unfortunately, Richard wouldn't listen to his calling and ignored the many small signs that were sent to him before his heart attack. He received physical signs, such as high blood pressure and ulcers. His business also sent him many signs, although he was too proud and too fearful to recognize them. His wife and children sent him many alerts as well, but to no avail.

*When you fail to listen to the signs, you will get bigger and bigger challenges until you learn to listen and follow.*

Often, these signs are physical manifestations in the body, such as symptoms or illnesses. These challenges are designed to put us into a present state of awareness. Richard chose to keep fighting in a business that he didn't love, and he avoided devoting his life to the talent and calling that he did love. For more than ten years, weight gain, insomnia, ulcers, and high blood pressure never fazed his ego-driven personality. Inside, his ego was literally killing him, and he just failed to see it.

So . . . What was the cause of his death?
Was it a weak heart and overexertion on the racquetball court? Stress at work? Was it poor eating habits that led to high cholesterol and blood pressure? Where should we place the blame? The physical component is

obvious. Of course we know that his talent and passion for roses was the calling that he ignored. All four dimensions of his "authentic being" played an equal and proportionate role in the termination of his body's capacity to function here on this earthly plane. Had Richard paid attention to the signs and symptoms at an earlier stage and honored his passion for roses instead of allowing his ego and his fears of the financial future to run his life, the outcome could have been completely different.

## HEALING VERSUS CURING

The adage, "you can lead a horse to water, but you can't make him drink" carries with it some very wise suggestions. In Richard's case, even though he was aware of the signs and symptoms, and even sought medical help, he feared the risk of change and chose to remain on the same path. He chose to row his boat not "gently down the stream" but angrily *against* the stream—which ended his life before it was "merrily, merrily, merrily, merrily, a dream." Richard chose the medical "Band-Aid" approach to *curing his symptoms,* through blood pressure medication, weight loss pills, and sleeping pills. This "cure" however, was only temporary.

Many people choose to take this "curative" route, which is the passive route to health. With this route, the health of the patient is in the doctor's hands and not the individual's. "Do what you can for me doc, I'm in your hands," were Richard's exact words. His belief, which is very common, was that if he paid the fee, it was up to the doctor to perform the service and deliver the outcome.

*InTuiTouch* defines this as the passive approach to health.
Healing, on the other hand, is an inside job. Once the person has been led to the water, he or she can choose to take on the responsibility to drink the water or not. This is known as the active approach.

*Healing is an ongoing process.*

Healing requires significant introspection, certain attitude, memory, and belief patterns, as well as a willingness to take risks and to let go of old belief patterns that create fears of our future and hold us back.

*You are never healed, but you are constantly in a process of healing.*

Healing is an ongoing, continuous, *active* process. The results are long-lasting because the power of healing comes from within. It is the connection to the authentic source of all that is. The God source within is the "*You*" behind the skin that you are in. Many people can understand this idea in concept, but their fears do not allow them to practice this in everyday life.

*True healing will not begin until you accept, from a deep knowing, that healing occurs from inside the self and not from the outside.*

If you are faced with a health challenge now, consider looking at it from this perspective.

# CHAPTER SEVENTEEN
## TAPPING INTO THE ENERGY THAT HEALS

# CHAPTER SEVENTEEN

# TAPPING INTO THE ENERGY THAT HEALS

A basic tenet of Albert Einstein's theory is that the universe is made up of matter, which is compressed energy, and that "energy is not created or destroyed, it simply changes form." A simple of example that supports this theory is the ice cube that is solid, then melts to a liquid, then may "disappear" into a gas vapor.

While it may appear that the physical makeup of the object has been destroyed, in reality, it has only changed form. If you think about it, the universe (God) represents pure energy; in fact, God is the *origin* of energy. If energy could be destroyed, where would the destroyed energy go? It can't go where the universe is not because the universe is all that there is.

Thus begins the understanding of energy as it relates to your health. In your everyday world, you have energy in and around you everywhere. Think about static electricity when you walk on carpet and touch metal and feel a shock. Or the balloon that magically sticks to the ceiling only after you rub it to create static electricity. Auric photography can capture the electromagnetic field around any living object and can verify moods and states of being. Kirlian photography has proven over the years that the "essence" of an amputated arm or the leaf that was cut will still show on the energetic field, even though the physical part is "gone." The image of the complete arm or the complete leaf remains intact.

A medical procedure known as Magnetic Resonance Imaging (MRI) captures an image in our body after separating the positive and negative electromagnetic ions of our body. Yes, our physical world is comprised of electromagnetic and electrochemical elements. The electrocardiogram or (EKG) measures the electrical potential in our heart muscle. The electroencephalograph or (EEG) measures the electrical potential within our brain cells. The cell walls of every organ and system in our body rely on the electrical potential to be balanced in order to complete their proper functions.

We have energy waves passing in and around us every day in the form of radio waves, microwaves, and television. We have light waves traveling through us in the form of lasers, ultraviolet and infrared. We have sound waves that travel through us in the form of radio, ultrasound, and infrasound.

We take so much of this energy for granted and accept its existence as "scientifically sound." Even though we can't see it, feel it, smell it, or touch it, we feel the *effects* of it every day in our lives. We know that we can objectify, quantify, and qualify its existence. Yet, when a five-thousand-year-old healing method known as acupuncture describes the "redirecting of energy through the use of needles," our scientific medical community balks at it, saying there is no proof of its efficacy.

When you enter a room and get an uncomfortable feeling about a person or a group of people, you describe this as a "bad vibe." Some people are more in tune with this ability than others, just as Mozart was a prodigy in music. Even so, those who are gifted in feeling these vibrations are often ridiculed as charlatans, witches, and quacks.

*The truth is that we all have the ability to sense the world around us, and what we are sensing is energy.*

## ENERGY AND HEALING

If you believe that healing comes from within but can't let go of the traditional, mainstream medical approach of leaving healing in the hands of the doctor, then you will have a conflict in your belief system. You will be essentially living in two worlds simultaneously. This will undoubtedly cause a loss of energy, a loss of certainty, and a loss of health. Imagine the amount of energy that is lost when you rest all of your faith in someone other than your own internal power. You are relinquishing your internal power (God) to another lower source (the doctor). Energetically, this is the equivalent of opening up a one-way valve to the outside source in hopes that the doctor can take care of you instead of you taking care of yourself. This type of passive curing will take away symptoms temporarily but at a very high energetic cost.

Healing from within requires an enormous expenditure of energy. Introspection and understanding require you to remain humble, open minded, and open-hearted while at the same time detaching from the outcome.

## TRUE ACTIVE HEALING HAS NO BOUNDARIES.

Even death is an option in the menu of infinite possibilities. We do not understand all of the reasons for outcomes, but we know that there is a divine design for all things at a higher consciousness level.

Those who live in both worlds (passive medical and active self-responsibility) expend the greatest amount of energy and produce the least favorable results. In the chapters on the chakras and energy, you learned how an infinite number of people, places, and events cause energetic losses beyond the physical dimension. Those who continuously relive the events of the past, or continuously live in fear of future events, draw much of their bio-energy away from their present state of being, which is where they need it in order to heal.

If you or someone you love is looking for a way to truly heal, remember, true healing is active, and it occurs from the inside. If you look for someone to heal you from the outside, it is the same thing as asking someone else to digest your food for you.

Another person could not do that even if he wanted to, because all of digestion occurs from the inside. We don't even consciously perform this everyday process. We must rely on a power greater than us to do it for us. This incredible daily process not only breaks down the food that we ingest, but it also distributes it to all the right places in all the right amounts at all the right times. This energy, this source, is our inside doctor, twenty-four hours per day, seven days per week.

## HOW TO TAP INTO THE ENERGY THAT HEALS

Accept that you are a four-dimensional being that includes a physical body and an energetic body.

Try to identify who or what event in your life is taking away your energy and preventing you from staying in the present.

Identify where you are losing a lot of your energy and where you are gaining a lot of energy.

Ask yourself why you are allowing your energy to be lost (question #2).

List ten drawbacks to losing this energy and find ten benefits to you of losing this energy.

What is the lesson that the universe is trying to teach you with this experience?

Ask yourself: Am I willing to do what it takes, travel any distance, pay any price to reach an understanding of the "why" to my healing process?

By following these action steps, you can be sure to get healing underway.

Remember, you must remain congruent in your mind, body, emotions, and spirit. Do whatever feels right in your spirit, then discipline your mind and emotions to follow and honor your body with habits designed to enhance the healing process at any price.

# CHAPTER EIGHTEEN

## CAGED THINKING

# CHAPTER EIGHTEEN

# CAGED THINKING

There was a baby Bengal tiger named Benji who was placed in a 20 foot by 20 foot cage and sent to the Philadelphia Zoo. Eventually, the building of an enormous habitat would allow him to roam. While in the cage, though, Benji would pace three steps forward, turn to the right, then three steps forward and turn to the right until he would complete a full circle. As politics in the budget go, construction was temporarily halted due to lack of funding; so, the only exercise this beautiful creature could do was three steps forward, turn to the right and three steps forward, etc. Unfortunately, construction of Benji's enclosure was not completed for nearly three years! Finally, on the glorious opening day of the Bengal wildlife habitat, the cage door was lifted for young Benji to run free in his new home. The crowd's anticipation was brought to a stunning halt when they noticed that Benji didn't run free. Even without the bars in front of him, Benji continued to pace: three steps forward, turning to the right, three steps forward. Benji was confined to a world without freedom, and that was all he knew. To this day, Benji continues to pace in his imaginary cage, never experiencing the freedom of his new habitat because of his stunted perception of reality.

The bars in our mind are stronger than steel. How many limitations have we placed in front of us based on our *perception* of what is real? When a doctor tells a patient that he only has three months to live, how does he know that? How many people have defied the odds and continued to live for ten or twenty years more? No one really knows except the doctor within. Up until the last three hundred years, the doctors and healers of the community were the priests, shamans, and witch doctors. The spiritual aspect of a person was addressed as a principal factor of all illness. It wasn't until Rene Descartes, considered the father of modern medicine, put forth his radical viewpoint that separated science and religion, that the approach to healing changed. With the advent of laboratory sciences, Western medicine looked at the body as a machine separate from the spirit and separate from the mind. Like Benji the tiger, our society has been brainwashed into thinking that for every illness, there is a pill or a potion that will cure it. Such caged thinking has created a social dependency on Western doctors and a trillion-dollar pharmaceutical industry that creates a link between pharmacies, doctors, hospitals, and insurance companies.

It is only in the last fifty years and predominantly in the last twenty, that the medical approach to health care is being challenged by a philosophy called "alternative medicine." Modern medicine's failure to produce lasting results, combined with the increase in environmentally caused illnesses like cancer, has forced society to question the efficacy of medicine and its chemicals. The return to the philosophy of an integrated, holistic approach is being supported by the same sciences that condemned this approach. It is a movement going full circle.

## SPECIALIZATION AND COMPARTMENTALIZATION

Modern medicine has come a long way from compartmentalizing the human body, now recognizing that anything done to one organ or one system in the body deeply affects the whole body. If we remove the appendix simply because it *someday* may become infected, there is a price to pay within the body.

Each of us is a whole human body where each system and each organ co-depend and interrelate to each other. The same holds true for the link between our physical functions and our mind and spirit. To think that one is not divinely interconnected to the others is totally unnatural and sends us backward on the path of conscious evolution. Our emotions deeply affect our body. The obvious connections commonly seen are problems with blood pressure, heart rate, insomnia, rashes, and weight gain or loss. These symptoms are signals from nature that something is in imbalance. Everyone agrees that ignoring the warning signs could be dangerous. Masking the signs and symptoms through drugs (pain killers etc.), surgery (liposuction, appendectomy, etc.) or alcohol (drowning the problem) has never corrected the cause, and until the cause is acknowledged, the cause cannot be treated. It takes courage and deep introspection to acknowledge the truth of these connections, but, as they say: "The Truth will set you free!"

# INTUITOUCH

# CHAPTER NINETEEN

## THE MIND / BODY CONNECTION

# CHAPTER NINETEEN

# THE MIND-BODY CONNECTION

D o the emotions that you experience on a daily basis affect
your health on the physical level? Does the over-burdened,
stressed-out executive develop shoulder, neck, and upper-back
pains because he or she is "carrying the weight of the world on his or her
shoulders"? If a wife fears standing up for what she believes in, can she
actually develop varicose veins because of those fears? Are those "knots
in your stomach" related to the ulcers and gastritis that you have been
suffering from for all of those years?

The question should not be whether there is a connection between the
mind and the body. The question should be what is the connection and
how can I eliminate the suffering? Human psychology and psychiatry have
done wonders to relate emotional factors to your health. Many authorities
have tied a very high percentage of physical illnesses to a person's emotional
state. I say with certainty that *all* physical illness are connected to your
emotional state and *all* emotional states are related to your physical state.

The work of researchers like Dr. Candice Pert, a neurobiologist, illustrates
this evolution. Her discovery of neuropeptides proved that thoughts and
emotions will trigger chemical reactions that cause all sorts of physical
changes in the body. One thought can influence changes in your sweat
glands, your blood chemistry, and your organ function. The very first
neuropeptide that was discovered was endorphin, which is considered
a natural form of morphine. In an emergency situation, thoughts and
emotions can trigger enormous amounts of both adrenalin and endorphin,
which can allow an individual to perform unexpected feats of heroism, such
as lifting a car to free a child trapped beneath it. This fascinating discovery
has opened up a new branch of science called psychoneuroimmunology
or PNI. The underlying scientific theory now is that thoughts (emotions)
transform into matter. It seems that the most powerful of these emotions
include fear, anger, sadness, and happiness. The emotions produce the
neuropeptides, which influence the cells by affecting their size, shape,
and function. Appropriate expression of appropriate emotions allows

the particular cells to return to normal after the expressed emotion. If an emotion is repressed, the body chemistry is unnaturally altered, which may cause radical changes at the cellular level, leading to imbalances and disease.

From the metaphysical viewpoint, you know that every organ and tissue is related to a specific emotion and acts as warning sign so that we get a message. By acknowledging and accepting these signs, you are *releasing* instead of *holding on,* thus creating more harmony and balance with the mind-body connection. This connection keeps you consciously connected to the universe, which ultimately allows you to live and enjoy life through your mission and talents. This expression of your gifts and talents will lead you to what most people believe is the reason you are here on the planet in the first place—*to learn to love unconditionally,* which is the integration of all emotions.

***If the seeds of disease are found buried in the emotional and psychological groundwork, then so too are the keys to healing.***

Carl Jung was quoted as saying, "the paradox is that the wound is also the treasure." Examples of physical illnesses that are specifically related to psychological or mental connections are infinite. Below is a small selection of the many examples of the mind-body connection and the associated perception of what this signifies.

| PHYSICAL | EMOTION | PERCEPTION |
|---|---|---|
| Shoulder pain | Responsibility/guilt | Carrying the Weight of the world |
| Knee pain | Flexibility in life | Can't move ahead |
| Arthritis | Holding on/ letting go | Life slipping away |
| Diabetes | Pessimism | Sweetness of life gone |
| Heart attack | Inability to love | Emotions bottled up |
| Ulcers | Self-destructive/ anger | Something burns me up; I can't stomach it |

Table—1

While these small examples outline only a few of the metaphysical connections, it is important to understand that these connections exist for *every* physical condition and <u>every</u> emotion. (Look for my book, *Heal Your Body with Metaphysics,* that expands *extensively* the mind-body associations.)

The more we learn about our health and how our health challenges are interrelated, the more we will understand how to allow the process of healing to occur. As Dr. Page so eloquently states, "For healing to occur, we must come to see that we are not so much responsible for our illnesses as we are responsible to them."

## REAL-LIFE STORY

*Ted's Story*

Ted was a forty-two-year-old father who, together with his seven-year-old son, left on a "father-son bonding" train trip, coast to coast, from New Jersey to California to visit some family members. On their return home to the East Coast, they received devastating news. During their absence, Ted's wife (and the mother of his son) had arranged to move in with her lover. The emotional trauma left the young son completely in turmoil and left Ted heartbroken. Within six months, Ted had remarried and started another family. Within eighteen months, a second baby was due. The "fire to frying pan" overreaction from the trauma left Ted with too much to handle. Within two years of this event, Ted experienced a heart attack while driving and died instantly. While the autopsy confirmed heart attack, I am certain the accurate diagnosis was really heart*break*. While there are always two sides to every story, Ted's perception of his being the victim created a constant energetic leak that literally ruled his life and eventually drained his heart. I know this to be true as Ted was my brother.

# CHAPTER TWENTY

## EMOTIONS AND PERCEPTIONS

# CHAPTER TWENTY

## EMOTIONS AND PERCEPTIONS

Recognizing who and what is draining you is the first step in retrieving your energy. Each of the seven chakras has specific emotional ties, both positive and negative, to it. As we are well aware, there are countless emotions within the human experience. Any emotion, energy in motion, is quite simply an exaggeration or minimization of the truth." It is based on your perception of what is. Truth is unconditional in that it doesn't rely on your perception to be correct or valid. To say that New York City has more residents living within the city limits than Buffalo is a truth that carries no emotion in the statement. To say that New York is a better place to live is a perception tied to an emotion, and not a truth.

Happiness, sadness, infatuation, resentment, good and bad, right and wrong, are concepts whose center balance is unconditional truth or in other words—unconditional love.

Remember, what Jesus said: "I am the Truth and the Light; I am Love."

When we marry, we usually repeat the words of a promise of unconditional love stating for "richer *and* poorer, sickness *and* health, happiness *and* sadness."

These words are considered love without conditions—or real, true love. All others are considered illusions of love or in other words, *exaggerations and minimizations* of true love. In our daily lives, we experience these emotions based on our *perception* of what is real. The atomic bomb on Nagasaki was considered devastating to Japanese and a Godsend to Americans. While it killed many, it saved lives also. It all depends on one's perception. The truth about the atomic bomb, when void of emotion, is that the event killed many people *and* ended the war.

The overprotective mother whose child commits suicide blames herself for not allowing the child to experience hardships from which to grow. This overprotection could have left her child weak and venerable to life's ups and downs. On the other hand, there is the professional, super-mom, who is too busy earning a living to have time for her child, who likewise blames herself for her child's suicide. She believes that she was never there enough to help direct her child from going down the wrong track. Who is right and who is wrong? It's all about perception. Each mother would serve herself and the world better if she ended the emotional drain of guilt over past events, where a significant amount of energy of the second (one on one relationship) and fourth (heart) chakras are being needlessly lost. The loss of energy creates an inability to function successfully in all areas of life, including interaction with family members: mentally, in constant reminders of the past; spiritually, in the guilt of feeling responsible for the death of another; vocationally, in the difficulty to concentrate and be productive; socially, with friends and community members; financially, in lessening of self-value and self-worth; and ultimately, physically, through nervous or systematic breakdowns called illnesses. The sooner each of us begins to live in the present and avoid the emotional losses of energy of living in the past, the more productive, fulfilled lives we will live.

# ARCHETYPES

In her book, *Chakras and their Archetypes*, Ambika Wauters describes archetypes as "a universal projection of the collective thoughts and emotions of humanity, commonly called the collective unconscious." Just as actors and actresses play characters in order to display a desired emotion, life itself offers you the opportunity to see your own archetypes played out, acting as a mirror for your inner state and showing you stepping stones by which you can move forward on your path of self-development.

You undoubtedly play positive characters in your day as parent, teacher, worker, healer, friend, student, and many, many other roles. Similarly, every day you play negative roles as victim, thief, controller, swindler, criticizer, excuser, and many more whether you want to admit it or not. Without the role of a negative character, you could not appreciate the value of life's emotions. A beautiful sunny day is only as special as the gloomy rainy day. Without the opposite experiences of weather, you certainly would not appreciate the sun. As you continue to experience all of life's emotions, you will notice that you will play certain roles more often than others. These are your core archetypes that represent both your strengths and your weaknesses. They are in actuality, one and the same, just as a coin has two sides. Without one side or the other, the coin loses its value. Simply stated, emotions are the two sides of truth. Think about any emotion, positive or negative. Let us repeat the example of marriage. "Unconditional love" in a marriage requires the synthesis of loving a person "for richer and poorer, in sickness and health, in happiness and sadness." Accepting a marriage only during the good times is accepting marriage with *conditions*, and this is not unconditional love. Everything that you do in your daily life points you to one great truth: You are here on this planet to experience love in all that you do, in all that you have, and in all that you are. The synthesis of any two opposite emotions is love; and love is synonymous with truth. You need both positive *and* negative emotions to appreciate love in any of the experiences in your life. The archetypal roles, representing each and every emotion that you play every day allow you to experience all that life has to offer so that you master this truth called love.

# CHAPTER TWENTY ONE

## AWAKENING THE INNATE WISDOM OF THE BODY

# CHAPTER TWENTY-ONE

# AWAKENING THE INNATE WISDOM
# OF THE BODY

The process of healing is an unending string of events, a multiphase, multidimensional phenomenon. InTuiTouch is based on the premise that all healing occurs when the four dimensions of your authentic being are addressed simultaneously: physically, mentally, emotionally, and spiritually. Medicine has taken a mechanistic view of compartmentalizing illness into the various systems and organs of the body. InTuiTouch is a new approach that suggests that you open your heart and mind to the possibility that, by divine design, the health challenges that you encounter are here to help you and to teach you at a level that you may not yet comprehend. Once you acknowledge the possibility of such a lesson, you can learn to embrace your health challenge, honor it, and give it life as an entity with sound, color, smell, vibration, and texture. Then you can visualize not just the "what" part of it but the "why" part of it as well, so that you can then visualize the "how" part of the process of the healing to occur.

*This is only the beginning.* The real challenge comes from not only discovering your new innate healer within but also actually working with this entity from all four dimensions. It is an amazing process of spiritual growth for anyone who is willing to open up and do the work. This form of healing reconnects you with your authentic being in *totality*. It is the way we saw the world when we were born into the world. It is the way of healing. It is the only way to take full responsibility for your well being and health. Obviously, this form of thinking is new to some. It is an elevated state of consciousness that recognizes a divine design in everything—including illness. There is no one to blame and there are no coincidences that put you where you are in your life. Open your heart and mind, and enter into the world of the innate healer within. First and foremost, you must be grateful for the life you live.

What do I tell my patients when they ask me, "Doctor, how can I be grateful during this most troubling time?" I guide them out of their logical, analytical mind (left brain) and into their sensitive, creative, emotional mind (right brain) by doing a very simple exercise for seven consecutive days. Each morning upon arising and then again each night before going to sleep, make a list of ten reasons that you are grateful in the seven areas of your life (spiritual, mental, vocational, financial, familial, social, and physical). That totals seventy lines. Yes, that's right, seventy lines each day for seven days for a total of 490 reasons that you are thankful for being alive. Does it sound like a very difficult task? Henry Ford said, "If you think it is, or you think it isn't—you are right! Making this list may be time-consuming but it isn't difficult and certainly not impossible.

*After all, it is only your health and possibly your life that we are talking about here!*

By the end of seven days, you will come to a realization that your life has been filled with both crisis and blessings, sickness and health, tragedy and comedy, challenge and support. The recognition of this duality takes you out of your *reality* (illusion) and into *actuality* (truth). You are now rising above your paradox and into the sea of infinite possibilities where true healing occurs.

*Gratitude levitates . . . Ingratitude gravitates.*

# THREE QUESTIONS TO ASK THE INNATE HEALER WITHIN

1) *Do I believe that I have the right to be healthy?* Health is a birthright and the mere fact that you are here on this planet says you are included in this birthright!

2) *Can I accept the fact that there are no coincidences in life and that my health challenge is a blessing that I haven't seen yet?* Symptoms and ill health bring our awareness to the forefront and we become very "present."

3) *Do I understand the importance of finally honoring myself and my specific needs as much as I honor the needs of others around me?*

Think about this: how can you help others if you are sick and weak? Honor the balance of receiving as much as giving. It does wonders for the self-esteem and raises your personal value.

# CHAPTER TWENTY TWO

## IS OUR HEALTH DIVINELY DESIGNED?

# CHAPTER TWENTY-TWO

## IS OUR HEALTH DIVINELY DESIGNED?

For those who believe, no proof is necessary. For those who disbelieve, no proof is possible. Your perception of a divine order within worldly chaos is what matters, and this order seems to follow a universal law, meaning that it will withstand the test of time. The order of things was the same ten thousand years ago as it is today and will be for the next ten thousand years. This is what divine design is all about.

When astronaut James Lovell watched our planet from outer space for the first time, his perception of reality changed forever. He understood clearly that we humans are part of a much bigger picture. He saw the magnificent perfection and order surrounding him. While he was venturing to outer space, he remembered the wars and turmoil that he left back home on earth. What he witnessed from the perspective of outer space, however, was a graceful, slowly orbiting planet among the other celestial bodies. From this viewpoint above the chaos, he was able to identify the peaceful tranquility that existed simultaneously on the same planet that he perceived as tumultuous and at war. He concluded that mankind is not in control of this planet, the solar system, and certainly not the universe. Humans are nothing more than inhabitants of spaceship earth, going along for the celestial ride. Even mother earth herself is an inhabitant of a bigger order that undoubtedly was divinely designed. From this perspective, we humans are just a small part of a bigger picture, just as the ants are a small part of our natural picture. Do humans control the changing of the seasons, hurricanes, or earthquakes? Do we "tame" Mother Nature or merely *interfere* (consciously or unconsciously) at every chance we get—then pay the price of our unnatural interference?

B.J. Palmer, author, philosopher, and the single most influential chiropractor who ever walked this planet, wrote one of the most profound commentaries relating to man's interference with nature and how chiropractors worldwide can take a stand on natural health care. All healers dedicated to natural health care can relate to this poetic writing, titled *The Truth*.

## *The Truth*

"We chiropractors work with the subtle substance of the soul. We release the imprisoned impulse, the tiny rivulets of life force that emanate from the mind and flow over the nerves to the cells and stir them into life. We deal with the majestic power that transforms common food into the living, loving, thinking clay that robes the earth with beauty and hues and scents the flowers with the glory of the air.

In the dim, dark, distant long-ago, when the sun first bowed to the morning star, this power spoke and there was life. It quickened the slime of the seas and the dust of the earth and drove the cell into union with its fellows in countless living forms. Through eons of time, it finned the fish and winged the bird and fanged the beast. Endlessly, it worked, evolving its form until it produced the crowning glory of them all. With tireless energy, it blows the bubble of each individual life and then . . . suddenly, silently, relentlessly, it dissolves the form and absorbs the spirit into itself again.

And yet you ask, 'Can chiropractors cure appendicitis or the flu?' Have you more faith in a spoonful of medicine than in the power that animates the living world?"

These words ring true today more than ever, as we embark on a new era of energy medicine.

The healing of a wound, the elevation of body temperature in response to an infection, and every bodily function in between is controlled by a power (energy) that was divinely designed. Some call it the infinite intelligence while others call it Mother Nature. Others simply call it God. By any name, we must recognize it for what it is, and allow it to humble us. When we finally release ourselves to this power, through humility and gratitude, miracles occur. It is then and only then that we step into that stream of consciousness that is perfect and our lives flow *with the stream* instead of against it. Why would anyone wish to flow against the stream? The answer is simple: EGO. The minute that we try to separate from the source (God), we separate ourselves from each other and begin to perceive a duality. We create our own world full of judgment of right and wrong, good and bad, better or worse, sick and healthy.

# CHAPTER TWENTY THREE

## YOUR DIVINE PURPOSE
## IS YOUR SPIRIT

# CHAPTER TWENTY-THREE

# YOUR DIVINE PURPOSE IS YOUR SPIRIT

The "authentic you" mentioned over and over again in this text comprises four dimensions: mind, body, emotions and . . . spirit. The word "spirit" can be used in many different contexts. In this book, I refer to spirit as the "etheric" part of you. You have a divine calling in life; your divine calling is composed of your mission, talent, and destiny as a human being.

Everyone has been given a talent in this life. Some play the piano, some excel in athletics, while others have been blessed with the ability to act. There are mechanically minded individuals and others who are analytical by nature. These are all talents. The question is, what are you doing with your talent? You have a mission and a destiny. The mission is the reason why your talent exists. You are divinely designed to be of unique service to others (the art of giving), which results in ultimately serving yourself simultaneously (the art of receiving). Destiny is the roadmap upon which the talent is unveiled. On the surface, it may appear that others have been blessed with *greater* talents than you. The fame and fortune of others in comparison with your own life may have left you with the perception that successful people have been given more talent. While everyone's talent varies, the quantity and the quality of blessings are equal in number to all. It is only that some people *acknowledge* their talents and *do* something with them, while others look for ways to avoid them because of the effort required of change or fear of the unknown outcomes or of their future.

Remember the quote from Robin Williams in the movie *Dead Poets Society?* He said, "The song of life will continue playing and you *will* contribute your verse, but the question remains: 'What will your verse be?'"

You do have a purpose that includes your unique mission, talent, and destiny. It was given to you to maximize the expression of the "authentic you," which is your soul. As Dr. Demartini explains in his book, *Sacred Journey,* the word SOUL is the acronym for *Spirit Of Unconditional Love.* Ultimately, your purpose is to express unconditional love to everyone and everything that you encounter while here in your life. When you are not expressing the authentic you, you are "off of your purpose." Many times, distractions in life will take us off purpose. These are the results of living in fear of the future or living with guilt over events in the past that seem to continue to haunt us. Both the fears of the future and the guilt of the past keep us distracted from being in the present. But the present is the only time we actually have, and it is the precise place where being "on purpose" can be demonstrated. Whenever you are living in fears of the future or guilt of the past, your physical body will begin to receive reminders that you are off purpose in an effort for you to return to being present and "on purpose." You experience those reminders in the physical expression of symptoms.

*Symptoms are reminders that you are off purpose and are designed to help steer you back to your purpose.*

Your mission drives you because of a purpose greater than yourself. Your talent directs you because of the uniqueness of your specific endeavor. Your destiny empowers you to follow through to do whatever your inner voice is calling you to do and have all that you deserve in your life. The formula is as follows:

MISSION + TALENT + DESTINY = PURPOSE

The *In Tui Touch* Method was created to give you a way of linking your *purpose* to your *mental* and *emotional* misperceptions as well as your *physical* symptoms, thus setting up a pattern of congruency between the *mind, body, emotions,* and *spirit,* which ultimately leads to healing in the true sense of the definition.

Remember the definition according to Dorland's medical dictionary:

*Health:* A state of optimum physical, mental, and social well-being, not merely the absence of symptoms or disease"

# CHAPTER TWENTY FOUR

## PUTTING IT ALL TOGETHER

# CHAPTER TWENTY-FOUR

## PUTTING IT ALL TOGETHER

The energy from the universe is running through you like a receiving and broadcasting station at all times. The truth of the matter is that the "matter" of which you are made of is the same energy that runs through the entire universe. The molecules of "you" are held together in this material world at a frequency and vibration that manifests itself as a human being. Your four parts (or dimensions): mind, body, emotions, and spirit, are also simply different vibrations that are resonating at different frequencies, which combine to make you unique. You perceive your physical body as uniquely physical, but in reality, it is only a vibratory wavelength. Our consciousness is considered merely a "cloud of electrically charged particles of light," according to Fermi Award winner Freeman J Dyson.

"Light," "love," and "God" are interchangeable words for the same thing: energy. You are here on this planet, expressed in a physical form of energy, for one reason and one reason only, and that is for you to experience love in its many forms, such as through your trials and tribulations as well as through your infatuations and exhilarations. We have heard the expression, "Through good times and bad, rich and poor, as well as *sickness and health.*" That has been the expression of unconditional love throughout the ages. You are now in a unique position, able to look at health and health challenges as a way of learning to love life and experience new ways of understanding what each challenge brings to you. Whenever you are ill or have experienced an "accident" that causes you physical, mental, or emotional pain and suffering, ask yourself, "What is the message that the universe is sending me and that I am trying to learn here? What is the connection metaphysically to my current health challenge? What lesson of love have I not been getting?"

"The quality of the questions that you ask determines the quality of the life that you lead." Look deeper into your challenges and know with certainty that they are divinely designed to help you on your spiritual journey of love and light. Remember: Love penetrates through barriers that thoughts can't even enter.

The message that God is giving to you in the form of a symptom or a health challenge is out of *love* not out of some form of punishment. God is Love and that love is *unconditional*. Punishment is an indication that God is unhappy with you and cannot forgive you for some mistake that you perceive that you have made while learning your life lessons here on this planet. Is that Love? Is that unconditional?

Is there really a God who will send you to suffer illnesses here on earth and then possibly send you to a place to suffer "eternal damnation" for something that you did or didn't do intentionally or unintentionally? While I am not here to change your *religious* beliefs, I am here to change your *health* beliefs so that you can get on with the process of healing yourself or others. The **InTuiTouch** way to health maintains the position that *"everything serves"* at a higher spiritual level, and the only way to elevate your vibratory level to a healthy level is to rid yourself of any illusions that your health challenge is somehow a punishment that you deserve. This is somehow derived from the illusion of guilt for something that you did or didn't do along your life path. Guilt and the subsequent feeling of deserving punishment vibrate at a very low frequency and rob your physical body of vital energy that maintains your health at the cellular level. Change your beliefs and you will change your vibratory level. Changing your vibratory level will undoubtedly change your health at the cellular level. Fear and guilt are exaggerations or minimizations of what is true. All illness is based on the illusion of seeing more negatives than positives, more guilt and fear than truth, which is love.

As we continue to consciously evolve in this age of Aquarius, the age of love and enlightenment, with powerful teachers like Dr. Demartini, as well as the other great authors and teachers mentioned in this book, we are coming to the realization that fear and guilt are just illusions that we have created to bring us to the equilibrated state of love. We are here to love. Our physiology is designed to be equilibrated, which is love in the physical form. Our minds are designed to see the Duality of life (balanced emotions) and all of the emotions that swing back and forth from good and bad, right and wrong, pretty and ugly, happiness and sadness, infatuation and resentment. The synthesis of which is again . . . love. Our spirit is designed to see both sides of unconditional love in the form of hope and hopelessness, of faith and atheism, even of the lower-minded form of love and its counterpart, hate. We call this balance *unconditional* love. We find this consistent throughout the universe. In chemistry, it is called neutralized charges; in astronomy, lightness and darkness; in physics, action and reaction; in religion, good and evil; in politics, war and peace; and yes, finally, in health, we call this homeostasis. There exists a duality in the universe at the macrocosmic level as well as the microcosmic level and . . .

## *Love*
## *is the synthesis of all that is.*

I ask that you only acknowledge that statement from your own experience (wisdom) and not from what others have told you (beliefs). Open your mind and your heart to this realization, and you will experience miracles beyond your wildest imaginations and dreams.

I'll see you there. I love you.

Dr. Jim

# SPECIAL DEDICATION

One of the signs of a truly great man or woman is that that person makes an impact on the world quietly, never looking for the recognition or applause of those he or she is serving. This is also fits the description of the person who is a truly great friend: one who gives out of love, expecting nothing in return. Such are the qualities of my friend Jim Seltzer. I am certain that he would not even approve of this special dedication in my book. Nonetheless, I am compelled to acknowledge his contributions to this book and to my life.

Without ever needing an explanation, Jim has offered me support through his wise insights and recommendations. He is a master of life. He is a true sage who sits outside of the limelight and enjoys the ride.

I will be forever grateful for his presence in my life and very proud to call him my friend.

Thank you Jim, you *are* like a brother to me and I love you.
Jim

# REFERENCES

Andrews, Ted. *Animal Wise*. Dragonhawk Publishing, 1999.

Chopra, Deepak. *The Seven Spiritual Secrets To Success*. Publishers Group West, 1993.

Demartini, John F. *Count Your Blessings*. Element Books, Inc., 1997.

Dyer, Wayne. *Manifest Your Destiny*. Harper Collins Publishing Company, 1997.

Gordon, Richard. *Quantum Touch*. North Atlantic Books, 1999.

Hill, Napoleon. *Think and Grow Rich*. Ballantine Publishing, 1960.

Jampolsky, Gerald. *Love is Letting Go of Fear*. Celestial Arts Publishing, 1979.

Leadbeater, C.W. *The Chakras*. Theosophical Publishing House, 1988.

Lindgren, C.E. *Capturing the Aura*. Blue Dolphin Publishing, 2000.

Myss, Caroline. *Anatomy of the Spirit*. Three Rivers Press, 1997.
Osho. *From Medication to Meditation*. C.W. Daniel Company, Ltd., 2004.

Walsch, Neale Donald. *Moments of Grace*. Hampton Roads Publishing Company, 2001.

Wauters, Ambika. *Chakras and Their Archetypes*. The Crossing Press, 1999.

Zukov, Gary. *The Seat of the Soul*. Free Press, 1990.

# Audio Cassette Tapes

Chopra, Deepak. *Synchro Destiny.* Nightingale/Conant Publishing/ Recording.

Demartini, John F. *Sacred Healing. 1993.*

*Quest . . . Energy, Power, and Spirit,* Volume 3. Simon and Shuster Sound Ideas, 1997.

# About the Author

**Dr. Jim Bourque Starr** opened his first chiropractic clinic in 1979 after graduating from the Los Angeles College of Chiropractic. He went on to successfully operate five clinics simultaneously throughout the central California coast. In his "darkest night of the soul," Dr. Starr received his most inspiring message, which led to the creation of the *In Tui Touch* Method and his book, *In Tui Touch—Healing Through the Gift of Intuition and the Art of Touch*. This book and the ensuing seminars and lectures are the evolution of this message. He now travels throughout the United States and Mexico, teaching and inspiring students from all walks of life about this wonderful and simple healing method. He has been interviewed on national TV and radio stations, and he has lectured at countless natural health centers and hospitals.

He has authored three other important books, *Heal Your Body with Metaphysics,* which serves as an informative guide into the mind-body-energetic and spiritual connection of the most common and difficult health challenges that we face today; *The Seven Myths of Nutrition and Wellness,* a simple, no-nonsense guide to proper health habits without the fanaticism of the health gurus, including straightforward truths about vitamins, minerals, and general supplementation and *The Redneck's Guide to Saving the Planet* - a humorous but serious reading for anyone who is interested in doing their part in going "Green."

For information on seminar dates, visit Dr. Jim at www.drjimstarr.com

# Introduction to the
# InTuiTouch Method

The information shared in this section of the book is designed to give the reader the ability to experience the **InTuiTouch Method** firsthand. The procedures that follow will help the reader promote the healing of self or the healing of others. It a step-by-step procedure that is so simple that one might think that there is something missing. Many people think precisely that, until they have experienced the results for themselves or those to whom they are applying the method. The procedures are taken directly out of the course manual for *InTuiTouch* Basic *Certification Course*, taught by Dr. Bourque Starr and his team at seminars across the United States and Mexico.

Readers may enjoy the straightforward approach and the clarity of explanations regarding positioning and application of this healing technique. The contents, however, cannot replace the value of the *two-day workshop* that leads to certification. This is a hands-on, intense course that includes the theoretical and practical application combined with a fun-filled mixture of guided meditations, consciousness-expanding topics, great friendships, and bonding with fellow healers and practitioners. Nothing can replace the value of this experience. Obviously, the energy shared in a group setting with enthusiastic students and the magical presentation of Dr. Jim Bourque Starr and his team of teachers is a weekend to remember. Many participants express going through a healthy version of an "identity crisis," as they are not the same person who left home for the weekend when they return home.

## *What you have before you is very special.*

**InTuiTouch** is unique. You can literally clear the pathway for the only true healing force that exists, which is the innate power that resides within each of us. Illness isn't about emotions, physical trauma, genetics, or pollution. It's about all of them together. Nothing is separate in this world, as you will see.

Whether you are lay person called here to heal, or a health-care practitioner looking for a better, clearer way to help your patients, you will have an opportunity today to discover and uncover, the underlying imbalances that rob us of our vital flow of energy, which ultimately results in disharmony, imbalance—dis-ease.

*Caution: Simplicity rules!* This is a straightforward, easy-to-learn method involving the synthesis of bio-energetic chakra work, the powerful healing effects of human touch, the recognition and acknowledgement of the emotional connections that are never separated from the physical, and the utilization of one's intuitive, spiritual, and natural gifts in a manner designed to participate in the healing process of another human being.

Thank you for being here and thank you for Be-ing.

## SEVEN ATTITUDINAL

### INGREDIENTS OF A HEALER

Everyone has the capacity to heal.
Healing is not sacred only to the chosen few. Somewhere along the lines of life, we have been fed the belief that healers are naturally born with a supernatural gift. This is a fallacy for two very important reasons.

### #1: The "Healer" doesn't do the healing.

Without the need to go into expansive explanations of the existence of an omnipotent power that is clearly beyond human comprehension, let us at least be humble enough to admit that a human being is only an instrument that assists in the healing of another. While this can ultimately be very important in the healing process, most of the time, it is not necessary for healing to occur. In fact, most healings and reparations go on in the human body without our awareness. Even the prodigal piano player needs to practice and develop his or her God-given talents. Likewise, healers come in every color, size, and shape and develop their talents as they go along. (That is why they call their business a "practice.") Everyone has the ability to develop healing skills to assist another in their healing process. While it is true that some may have more tendencies and motivations than others because healing may pique their interest, the skill is a learned skill, whether it is with the scalpel of a surgeon or the energy work of a Reiki master.

That is the good news because anyone reading this will understand that they too can and will learn the **InTuiTouch Method**.

#### #2: The "Healer" is a human being just like You and I.

While it is clear that we are born with certain talents and tendencies, every skill is learned and developed over time. As with any learned skill, there is some basic information that teachers must share with students during the learning journey.

### SEVEN ATTITUDINAL INGREDIENTS OF A HEALER

The basic attitudinal ingredients for anyone entering into the healing process of another include the following seven attributes:

> Love
> Empathy
> Detachment
> Giving
> Certainty
> Humility
> Gratitude

**Love**: Naturally, love is the key to any healing. In order to be an effective instrument that passes the healing energy of the universe (God) into another human being, the healer must love what he or she is doing as a vocation and love the entire process of healing for what it is. By loving the patient for who he or she is, without judgment of outward or inward appearances, the conscious effort will go toward the assistance of any other human interaction. Think about each cut and wound that self-heals without conscious interaction. How many colds have disappeared without pills or soups or injections? It happens all the time.

**Empathy:** There is a huge difference between sympathy and empathy toward a patient or client who is in need of a healing. Trying to understand what a patient is feeling by putting yourself in their shoes for a moment allows the healer a great insight into how to approach the patient every step of the way in the **InTuiTouch** healing. The choice of words, the movements, the touch, the sound of one's voice, all play a great role in the healing process. Sympathy, on the other hand, is "feeling sorry" for the patient in need of a healing. By recognizing that each of us is going through our own evolutionary process either consciously or unconsciously, we must remember that there exists a perfect order in God's plan, and this particular patient is experiencing this challenge to grow and learn to love. By remembering and understanding this concept, we can do away with the sympathetic response and instead, approach the patient with empathy.

**Detachment**: One of the crucial steps for a healer is to understand that the healer is not responsible for the healing. The healing is a result of the patient's connection to the greater power. The healer can assist in this process but ultimately cannot take credit for the healing. The concept of "take no credit, take no blame" is paramount in the healing process. The healer must remain separate and detached from the outcome of the healing. Anticipating a desired response and the resultant disappointment of not having achieved that result, only demonstrates how the human ego gets in the way of divine intervention, which is the source of all healing in the first place.

**Giving:** While it may appear to be a very simplistic idea, the art of giving truly must be recognized for its important role in the healing process. In our external world, we are constantly bombarded with the pressures of daily life. The economics of living, the family, and social pressures all can divert our focus from why we do what we do as healers. If we constantly remind ourselves of the universal law of giving simply for the sake of giving unconditionally, we know that we will receive back abundantly in ways we don't even recognize. Give from the heart, all that you can to your patients or clients, and the healing process will be greatly enhanced.

**Certainty:** As in all aspects of life, he who has the most certainty rules, and he who has the less certainty is ruled.
Healing requires you, the healer, to understand this law, especially when a patient is ill and in need of your help. Confidence gives the patient a sense of calming and serenity that naturally helps the body and mind in the healing process. Whenever you as the healer are in doubt, simply revert to your basic understanding of where healing originates. The power that made the body (God) is the power that heals the body. With this knowledge of the origin of healing, there should be no reason at all that you, as the healer, cannot maintain the certainty needed to assure the patient that he or she is in the best of hands. After all, how could one be in better hands than God's?

**Humility:** As you can see, healing requires one to recognize that the source of the power to heal does not originate from human hands. All of us, however, have the divine right to be healthy and to be able to heal. To understand the magnitude of this great gift of healing, one simply has to recognize that we are part of a magnificent plan, a divine order. Maintaining a sense of humility keeps our ego under control; we know that *ego* stands for Etching God Out. Whenever we think that we are the source of healing instead of this divine power, we lose our humility and our ability to participate in the healing process.
Maintain your humility as a servant to your patients; remember that it is an honor for you to be able to participate in the process of healing yourself or others.

**Gratitude:** Each and every one of the above ingredients in healing ties in to this last attribute. Be thankful for life itself and for your ability to experience all that you do each day. This is something that all of us should remind ourselves about each and every day. As a healer, it is imperative to remain grounded and thankful for having the opportunity to be involved in the healing process of self or others. Gratitude opens the heart and allows the infinite source of light and love to enter our hearts, which opens the door for the healing process to occur.

These are the basic ingredients that give the healer the ability to connect with the source so as to serve humanity in the healing process. Keep these principles in your mind, but more importantly, practice and live these principles each and every day. Your impact on the world will be greatly enhanced when you do.

## NATURAL LAWS OF ENERGY

Working with the flow of natural energy and natural processes allow the healer to tap into the most powerful healing tool ever discovered. It is available in an abundant, never-ending supply and available to everyone who has an inclination or desire to learn how to capture it, not just a chosen few. In the **InTuiTouch Method**, you will hold energy in your hands and direct it intuitively to the areas of the body where the energy is needed. This may not be the same location that you or your patient thinks that the energy is needed, but it will be the location that the universal healer within us knows that energy is needed.

Intuition is a gift that has been inherently received to help guide you along your path and to support you as an adjunct to your five senses in your daily life.

Remember: The power that made the body heals the body.

The intuitive part of The **InTuiTouch Method** has been given to you as part of your natural-born rights.

## EXPERIENCING THE FORCE

Without shaking, cleaning, pointing toward the
North, taking off your shoes, or wearing white clothes, simply do the following:

Put your hands in a relaxed position, with your hands facing each other about three inches apart while you are just sitting there.

Relax your mind and don't try to think what you are supposed to feel. Simply be.

After thirty seconds of this exercise, write down what you experienced—not necessarily what you felt. This does not mean that you could not have felt something, but it does mean that you did not have to feel something.

After thirty seconds, it was (very clear) (not very clear) that I:
Felt

Visualized

Smelled

Heard

Tasted

## Subtle changes produce not so subtle results.

Imagine a power or force that appears to us to be so light, so non-forceful—and yet leaves us with changes in our health status and ultimately in our lives that were not attained through traditional methodologies.

How can that be?

The moment that you think that you are making the change, your ego is in the way already. Drop this notion and let Mother Nature (universal intelligence) do her thing. Observe the response, don't anticipate it.

If you or your patient is experiencing health challenges at this time, examine how you think this health issue first appeared in the body.

Let's take a peptic ulcer, for example.

Throw away the idea that the cause is some virus or bacteria in the lining of the stomach (a common belief among diagnosticians at the moment), and let's get to the source. We know that energetically, the stomach represents the area of the third chakra. This is the center of the solar plexus and the center of our self-esteem and personal value. Our ego is centered in this area as well. If, over a prolonged period of time, we continue to "beat ourselves up" with emotional self-defeatism due to low self-esteem, the negative energy in this area of the body will vibrate at less than optimum levels (bad vibes) and, as a result, the cells of the organs and tissues of this area of the body (stomach) will not receive the proper amount of vital energy. Therefore, they will experience an imbalance of proper function. An overproduction of emotions, namely anger or the stimulation of beating oneself up, or low self-esteem, can cause an imbalance in the acid-base balance physiologically over a prolonged period of time. At first, the symptoms begin as acid indigestion and gastritis. When the antacids stop giving the temporary relief that they are designed to give, a more drastic medicine is prescribed (Tagamet), and more intense medical studies (endoscopy, GI studies, barium) are undergone to reach the diagnosis of "peptic ulcer." The plethora of medical intervention at this point is aimed at crisis care to get the symptoms under control so as to avoid a total collapse of the digestive system (cancer of the stomach, perforated bowel, etc.) Without medical intervention at this point, the entire organism (person) could self-destruct (death).

The point is that this level of disease in the stomach did not happen overnight. It took time and a series of very subtle losses of energy until eventually this level of deterioration was reached.

## THE POINT

The same subtleties at the energetic level that caused the losses of energy in the first place can be reversed at the energetic level, which ultimately, over time, will lead to changes at the physical level *provided that the four dimensions are addressed and balanced energetically.*

What do you think . . . that Mother Nature (God) cannot reverse the process that began in the first place? How dare you believe that the power that made the body (and the entire universe for that matter) cannot heal the body at any moment or stage of an illness?

THE NEW ERA
IN HEALING and
MEDICINE

*The paradigm shift is unstoppable.* If you stop for a moment and look at where Western science has come from and where it is going, you can get a better idea of how evolved this healing method called **InTuiTouch** really is.

- Leeches and potions
- Tinctures and herbs
- The grossest of gross surgery
- Chemicals and needles
- Electric current
- Radio and heat waves
- Microscopic
- Ultrasound, infrared, ultraviolet, and magnetic resonance
- Laser and light therapy

*And now we have evolved to* integrative, energetic medicine with **InTuiTouch** leading the way.

## THREE KEYS FOR HEALERS

### KEY NUMBER ONE

REMAIN HUMBLE TO THE SOURCE
The greatest key to receiving the gift of healing is the healer's ability to understand that he or she is not the one doing the healing. As I have said previously, the healer needs to get out of the way and let the universal source do whatever is divinely designed.

### KEY NUMBER TWO

NEVER RELY ON THE OUTCOME
Remember sometimes our best intentions are not in line with the best interests of a patient or client and certainly not always in line with the divine plan. Do not anticipate a certain response to the healing event. The **InTuiTouch** method is designed to balance a person in the four dimensions; after that, it is in God's hands.

Let go, let God! Give "thanks" where "thanks" is due, and leave it at that. Many healers understood this concept. None was greater than Jesus Christ.

*Think about other ways that we can become better healers.*

*Remember: Results are never guaranteed but are always appreciated.*

Wisdom = The instantaneous recognition that one's crisis is one's blessing.

## KEY NUMBER THREE

TAKE NO CREDIT, TAKE NO BLAME
One of the most difficult roadblocks in practicing a healing method of any kind is allowing your emotions to bathe your ego. Carrying blame and guilt for less-than-desired results of a healing is also counter-productive. It is not always easy to do, but it is something that every healer can try to be conscious of. You must understand that you can only do what you do. Try to remain neutral from an emotional standpoint.

# The four phases of review are as follows:

Rejection • Challenge • Qualification • Acceptance

At first, the theory is *rejected* as crazy and outlandish, even without any depth of understanding what it is about.
It is then *challenged* and determined to be "unscientific."
It is then tested and studied and finally *proven* as quantifiable and therefore *qualified*.
After such intense scrutiny, it is finally *accepted* as "scientific."

And yet, a five-thousand-year-old healing approach based on non-Western methods (acupuncture), is sometimes called quackery because the philosophy is based on energy patterns of flow (meridians) that have not been proven though results can consistently be reproduced. Chiropractic as well has long been dismissed as unscientific and quackery because its philosophical origin recognizes the innate intelligence within each and every one of us as the source of healing.

To date, there is no medical machine that can qualify or quantify the innate wisdom (universal intelligence) that continues to operate and control the functions of our human body.

In order to fully understand how energy works within your body, you should try to grasp how the universe supplies you with energy.
If you can understand a little more clearly where energy comes from and understand a little more of how the universe works, you will be closer to mastering the art called healing.

# INTRODUCTION TO THE
## *In Tui Touch METHOD*

Can you see why it was so important to understand the nature of energy and the connection with emotions before beginning the hands-on portion of the **In Tui Touch** Method?

Begin the actual technique by reconnecting with your *purpose.*

Understand the value of *your gifted hands.*
 *gratitude and humility* open the heart to the healing source.

Although we may have constructed a space station, we cannot create outer space. The source is the same life force that we are working with when you use the **In Tui Touch** Method. It is so real, so powerful, and yet so *simple.*

While there are many techniques that address the different aspects of healing, **In Tui Touch** addresses the *mind-body-spirit-emotions* in an unprecedented way.

Modern and ancient philosophers and sages agree that knowledge comes from the one, is spread by the few, received by the many, to return to the one.

*All healing points us to God (love) and leads us on our path, our mission, and our destiny.*

# TAKE YOUR HEALING SERIOUSLY

Always recognize the divine nature of why a client or patient has come to you. Be totally focused. Be in the present, your path. Look for the energetic imbalances through your intuition. Remove all doubts and fears and anxieties that you may have been harboring. In short, let your feelings "feel."

Remember: Changes will be made.
You now have the tools.

*Information + Exformation = Transformation*

# PRACTICING *InTuiTouch*

The key ingredient to any healing technique is *intention*.

**RELAX.** Put your client at ease. The more relaxed both the healer and the "*healee*" are, the better the flow of energy will be.

**BREATHE . . .** *Slowly and fully.*
*THREE* **STAGES OF BREATHING**

Abdominal

Chest

Apices

Breathing helps restore the flow of energy into our body.

**OBSERVE.** Intuitively get to know this person. Watch his or her body movements. Smile and send love to him/her. Listen and observe.

**ACCEPT.** Accept your client for who he/she is. Cast no judgment on his or her appearances or complaints about life, etc.

**REMEMBER:** You are here for the experiences in your life and they are here on this planet for theirs.

THE PHYSICAL DIMENSION
OF **InTuiTouch**

## STEP ONE: INSPECTION

*Front to back*—In this position, check the levels of the shoulders, hips, head tilt and any other imbalance that is obvious. Refer to your mini-guide, *Inspection Guide for InTuiTouch Practitioners,* if you are stuck on a finding that you cannot identify. At every point in the inspection process, pay attention to intuitive information that you may be receiving in any form that it may come.

*Side to side*—Try to notice any forward leaning of the head upon the neck. Any rounded shoulders should be noted as well. Look for excessive curvatures and the find the center of gravity throughout the body. Ask the body what it wants to tell you. Use your intuition!

*Back to front*—Check for any obvious imbalances of musculature, clothing lengths, and posture. It is a good way to double-check the findings of "Front to back."

Refer to the *Inspection Guide for InTuiTouch Practitioners* for further insight as to the mind-body connections, i.e., leaning to the right could indicate that the client is leaning away from something or leaning toward a decision, etc.? Again, use your intuition!

## STEP TWO: APPLICATION

*Energize*—In order to prepare for energizing, one needs neutralize one's thoughts of outcomes and sympathetic concerns for the patient's reasons for coming for a healing session.

Remember: There is no need to point to the north, wear a crystal around your neck, put on your white healing robe, or squirt sage essence all over the room. You simply have to *be*.

*Zoning*—This phase refers to reviewing the auric levels just outside of the body. Be aware of those levels as you enter into the space of your client. Do not underestimate these layers!

1) Relax
2) Breathe
3) Feel the energy as you pass through the chakra levels.
4) Look and listen intuitively for any messages you may receive during the zoning experience.

*Contacting*—This phase involves the physical contact of the body. The two levels of physical touch are as follows:

SUPERFICIAL = everyday issues
DEEP = core issues

1) Begin with the Para-spinal musculature, from the base of the skull down to the buttocks. Note where there is tightness and ascertain if it is one-sided or bilateral. Connect the chakra level with the musculature involved.

2) Feel the vertebrae, beginning from C-1 all the way down to L-5. When you lightly touch these bones, the body will "speak" to you as you practice and improve with time. Look and listen. If you feel nothing different, relax and don't force anything. You are still doing great!

*Intuiting*—This is done consciously throughout the entire **InTuiTouch** Method. Remain open to allowing your hands to "talk" to you.

Remember" Intuition comes through visual (clairvoyant), auditory (clairaudient), and touch (clairsentient) perception.

Your way will become clearer as you practice.

As a gentle reminder: Do not attach to the outcome.

## STEP THREE: SHARING THE ENERGY

You have now come to the highlight of the **InTuiTouch** Method healing session. This is where you have the most impact—because you don't do anything!

I mean that jokingly, but to a great extent, it is true. Remember, it is here that you are going to gather the energy and share it with your client (or "self," if your client is you).

1) The ideal position at this moment is to have the client lying supine (face up) on the table.

2) Have the client relax and breathe with eyes closed.

3) Tell him or her to let go of expectations of the experience and to just let the mind drift and relax.

4) Stand at the head of the table, extending your hands over the client's head and gather the energy as learned. Move your hands in a "gathering motion" with ever-increasing distances between your hands.

5) Pass your hands in the direction of specific chakras that you have identified, and continuously watch for signs of energy movement in all areas. Do not rush from one area to another. Let your intuition (and the client's body) tell you when to move on.

6) As in step 5, pass your hands (while gathering energy) to any place on the body without judgment as to why and let the energy continuously move through your hands.

7) When your intuition and the clients body signs have indicated that the session is over, the assessment and the physical dimension stage of the **InTuiTouch** Method healing session is almost complete.

8) Informing the client of the end of session—After you have finished with the healing session, gently and calmly tell your client that "at any time you would like to open your eyes, it will be fine." Usually, clients will move very slowly and peacefully back into the world.

9) Subjective reporting of the experience—As the client reconnects with you, you can ask him or her to describe the experience. *It is important not to try to lead the client by making suggestions of what the experience could have been.* The client is usually very quick to describe the event. Sometimes the client will remember nothing because they simply "left the building." Remember—every answer is okay!

10) Objective reporting of the experience—At this time, you have the chance to describe what you witnessed during the healing session. Describe body movements, breathing patterns, crying, or anything else that you found. Share with the client any intuitive messages or visuals that you received as well. All of this will serve him or her during the next forty-eight hours to three weeks.

Advise your client to pay attention and report any changes at the next visit, if there is a next visit.

11) Give gratitude to the power.

REVIEW OF THE PHYSICAL DIMENSION OF THE *InTuiTouch* METHOD

**Step one**: Inspection

**Step two**: Application

**Step three**: Sharing the energy

# ENERGETIC CHAKRAS

**CROWN VIOLET**
Grace: peace: faith
I enjoy peace and communion with God.
Fatigue; light sensitive; faith; depressed

**CROWN INDIGO**
Star: clarity: intuition
I think with wisdom and clarity.
Neurological diseases; blind; deaf; mental; seizures

**THROAT TURQUOISE**
Pyramid: truth: voice
I speak my truth and choose my highest good.
Throat; thyroid; TMJ; teeth; glands

**HEART PINK**
Heart: love: forgiving
I am gentle, loving, and open hearted.
Heart; lungs; breast; shoulder; upper back; /allergies

**GUT GOLD**
Sphere: self-esteem: self-worth
Who and what I am is enough. I am worth it.
Ulcers; digestion; liver; adrenal; arthritis

**SACRAL ORANGE**
Pyramid: power, sex, money

I deserve abundance for the right reasons.
Lower back; pelvic; infertility; kidney

**ROOT RED CUBE**
Tribe: grounded: support
I am secure and connected to all who love me.
Lower back; sciatica; immune; varicose; rectal

# SUGGESTED READING LIST

DEEPAK CHOPRA *QUANTUM HEALING* BANTAM 1989

DR. JIM BOURQUE STARR *HEAL YOUR BODY WITH METAPHYSICS*

BERNIE SIEGEL *LOVE, MEDICINE & MIRACLES* HARPER & ROW

DR. ALAN WATKINS *MIND/BODY MEDICINE* CHURCHILL—LIVINGSTONE

DR CHRISTINE PAGE *MIND BODY SPIRIT WORKBOOK* C-W-DANIEL CO.

RICHARD GORDON *YOUR HEALING HANDS* WINGBOW PRESS

CAROLYN MYSS *ANATOMY OF AN ILLNESS* BANTOM 1996

CAROLYN MYSS *WHY PEOPLE DON'T HEAL . . .* BANTOM 1998

DR JOHN DEMARTINI *COUNT YOUR BLESSINGS* PENGUIN BOOKS

RICHARD GERBER *VIBRATIONAL HEALING* BEAR & CO 1988

MARC IAN BARASCH *THE HEALING PATH* ARKANA 1995

NANCY ROSANOFF *THE INTUITION WORKBOOK* ASIAN PUBLISH.

LARRY DOSSEY *HEALING WORDS* HARPER COLLINS

AMBIKAWATERS *CHAKRAS AND THEIR ARCHETYPES* CROSSING PRESS

LOUISE HAY *HEAL YOUR BODY* HAYHOUSE INC.

BARBARA BRENNAN *HANDS OF LIGHT* BANTOM 1988

DAVID SMITH *HEALING JOURNEY* SIERRA BOOKS

LINDA MITCHELL *CHOOSE CHANGE . . . BEFORE CHANGE CHOOSES YOU*
IUNIVERSE

# RESOURCES

For a list of seminars and workshops in your area, or to purchase a book, video or audio recording, contact Dr. Jim Bourque Starr:

**Website:** www.drjimstarr.com

**Email:** info@drjimstarr.com

Books by Dr. Jim Bourque Starr

*Heal your body with Metaphysics.* The "Why?" behind your symptoms—At this very moment in mankind's history, there is a tremendous need for a book that will help teach us about reasons we suffer from the different illnesses and symptoms common today. We are in the age of a "consciousness expansion," and our understanding of the bigger picture, of our mission, our path, our destiny. *Heal Your Body with Metaphysics* is really both a book and a dictionary because it contains much more than simple definitions. It includes an array of information including the levels of chakra involvement as well the probable significance of mind-body connections of each of the more common illnesses from the simple cold to cancer. Intuition plays a big part in our personal understanding of an illness and there is plenty of room for personal interpretation.

*The Seven Myths of Nutrition and Wellness.* This simple, easy-to-read book contains secrets of living a healthy and fun filled life without the usual "fluff" of miracle cures and potions that fill the shelves of bookstores today. Dr. Jim explodes the myths of allopathic medicine and clearly puts in its appropriate category in the health care field: Sick care, not health care. Easy-to-follow instructions on exercise, vitamins, and proper diet are contained throughout the book, which concludes with a chapter on vitamins and minerals that informs you in very simple terms of the differences between them.

*The Redneck's Guide to Saving the Planet* - This humorous but serious book is an exploration into you can do to make a contribution to saving our planet and going "Green".

# InTuiTouch METHOD

## PERSONAL EXPERIENCE FORM

**SUBJECTIVE** (client/patient):

Did you feel any changes during your InTuiTouch Session? Y / N
What part(s) of your body did you feel the changes?

Head Area

Torso Area

Back Area

Other Area

What was the sensation that you experienced? Tingling? Temperature changes? (hot cold) Pains? Movements of air? Movements of Body Parts? Sense of Calm? Sense of Fear? Changes in breathing pattern? Visions? Voices?

**OBJECTIVE FINDINGS (InTuiTouch** Practitioner):

During the **InTuiTouch** session, the practitioner observed the following:

Statement of authenticity and permission to reprint or share this information:

My experience with the **InTuiTouch Method** healing session was as follows:

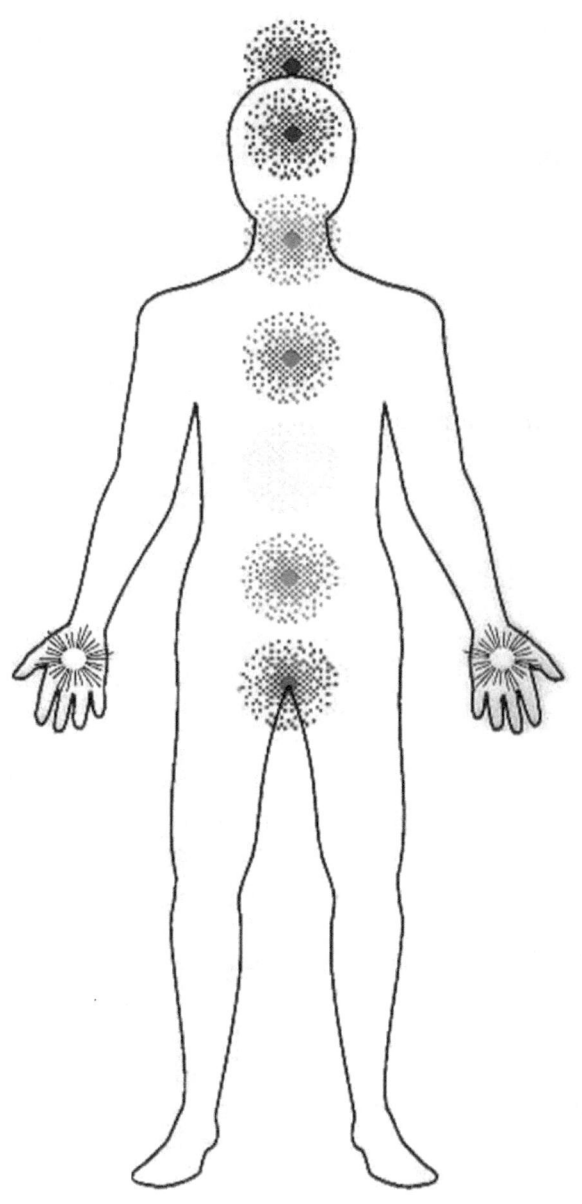